353.008 S372s 71-3494
Schooler, Dean,
Science, scientists, and public policy

0 0005 0038586 1

DISCARDED

Q   Schooler, Dean
127
.U6   Science, scientists, and
S28   public policy
1971
353.008 S372s

KCK Comm. Jr. College Library
727 Minnesota Ave.
Kansas City, Kansas 66101

Science, Scientists, and Public Policy

Dean Schooler, Jr.

 THE FREE PRESS, New York
Collier–Macmillan Limited, London

Science, Scientists, and Public Policy

Copyright © 1971 by The Free Press
A Division of The Macmillan Company

Printed in the United States of America

All rights reserved. No part of this book may be reproduced or transmitted in any form or by any means, electronic or mechanical, including photocopying, recording, or by any information storage and retrieval system, without permission in writing from the Publisher.

The Free Press
A Division of The Macmillan Company
866 Third Avenue, New York, New York 10022

Collier-Macmillan Canada Ltd., Toronto, Ontario

Library of Congress Catalog Card Number: 70-122274

printing number
1 2 3 4 5 6 7 8 9 10

Acknowledgments

Many hours and the efforts of many friends, colleagues, and teachers have gone into this work. Yet no matter how long an author's hours, the bulk of credit and gratitude must go to these others who contributed their time, thoughts, confidence, and friendship.

Major credit must go to Professors James A. Robinson, Randall Ripley, David Spitz, and Myron Q. Hale at The Ohio State University. Chapters 1 through 7 are drawn from a dissertation completed in 1969 at Ohio State. Some material, however, has been eliminated, and some segments have been substantially revised.

Dr. Robinson served as my adviser on the dissertation and should be associated most closely with the work as it now stands. The original inquiry and motivation developed in a seminar on Elites and Democratic Theory with Dr. Spitz, who is now at the City University of New York, Hunter College. Dr. Ripley contributed invaluable criticism and advice on the framework I have used to structure my thoughts and analysis. Dr. Hale (now at Purdue University) gave guidance throughout my doctoral studies.

Valued comment and criticism on earlier drafts came from Professor Charles O. Jones (now at the University of Pittsburgh), Carol Goss, Don Wyatt, and colleagues in the Department of Government at the

vi Acknowledgments

University of Arizona; Bonnie Steele, Al Damico, Tom Rosetti, and Harry Basehart at Ohio State; Professors John C. Burnham in the Department of History and Hershel Hausman in the Department of Physics at Ohio State; and Dr. Lewis E. Smith of Coshocton, Ohio. Typists are the backbone of the knowledge industry. For performing their lonely task, gratitude to Leta Hampton, Barbara Harrison, Janice Croft, and Kay Neves. More general acknowledgments are directed to my parents, G. A. Stewart, the Mershon Center for Education in National Security at Ohio State, and the Department of Government at the University of Arizona.

There remains one expression of gratitude. A man must eat, love, and be loved. And so Flosi was around throughout it all, being herself. Heather and Matthew joined us eventually, lending their hands to the effort.

The material in this book forms the basis for a course in Science and Public Policy (Government 286) now given at the University of Arizona. Segments and ideas, long since obscured by revision, appeared in a paper presented to the Graduate Student Panels at the Midwest Political Science Association in Chicago, May 2–4, 1968.

The National Science Foundation indirectly supported the work through a graduate fellowship from August 1966 through August 1968. However, no responsibility for the analysis or its conclusions rests with the Foundation.

No responsibility should be borne by anyone heretofore mentioned. Guilt by association must not plague a community of scholars, no more than it should plague a community of men. Collectively we can share any applause, but brickbats and rotten fruit must be targeted on me alone, all alone.

> Dean Schooler, Jr.
> Tucson, Arizona, and
>   Perryton, Ohio
>
> Assistant Professor
> Department of Government
> University of Arizona

Contents

v	ACKNOWLEDGMENTS
ix	LIST OF TABLES
xiii	INTRODUCTION

PART ONE. Science, Scientists, and American National Policy: The Post-War Years, 1945–1968

3	CHAPTER 1. Knowledge, Politics, and Power in the Executive Branch
25	CHAPTER 2. The Analysis: Framework and Definitions
63	CHAPTER 3. Scientists in the Executive Branch: Influence and Policy Arenas
73	CHAPTER 4. The Policy-making Process and Low Levels of Scientist Influence
137	CHAPTER 5. The Policy-making Process and Moderate Levels of Scientist Influence
167	CHAPTER 6. The Policy-making Process and High Levels of Scientist Influence
233	CHAPTER 7. Science, Scientists, and National Policy Making

PART TWO. Science, Scientists, and Public Policy Making: The 1970s and the Year 2000

257 CHAPTER 8. The Future of Scientists, Science, and Policy Making: Issues, Tensions, and Prospects

297 BIBLIOGRAPHY

317 INDEX

List of Tables

9	TABLE 1. Policy Types, Policy Processes, and Policy Arenas
14	TABLE 2. Selected Major Works on Science, Scientists, Scientific–Technological Elites, and Political Power
36	TABLE 3. Summary of Policy Types, Processes, and Arenas
58	TABLE 4. Summary of Factors Potentially Related to Scientists' Influence
64	TABLE 5. Policy Types by Policy Arena and Scientists' Level of Influence in the Executive Branch

Nam et ipsa scientia potestas est.
Knowledge itself is power.
   Francis Bacon
   Religious Meditations
     and
   Of Heresies

Introduction

This book deals with the relationship between scientists, science, and public policy making. In meeting that objective, the book sets forth the knowledge that scholars and observers now possess about scientists and public policy. As such, it is a "textbook" summarizing and synthesizing existing research, previous analyses, and observers' studies and theories. But it goes beyond the summary of a "text" and involves an analysis and research effort that should add to our existing stock of knowledge. Students benefit from a text, scholars from the research, and interested citizens and policy-makers from the analysis and discussion.

The book begins with an empirical analysis or research effort (Part I, Chapters 1 through 7). Most of the notable relationships between scientists and policy making in the United States have developed on a national level in the Executive Branch during the postwar period, 1945–1968. That is "where the action has been." The analysis, working from a selected group of policies, discusses those *factors* that have made scientists influential over policy and policy making. A wide variety of twenty policies are systematically studied, ranging from science policy to foreign aid policy and from defense policy to social policy. Chapter 7 discusses the findings of the research, draws some conclusions, and sets forth some implications. Basic studies and previous analyses are used and referred to throughout Part I.

But not all scientists' relationships with policy making have occurred in Washington and the federal government. Nor have they all taken place in the Executive Branch. What about new issues emerging between science, scientists, and policy making? What tensions can we expect to arise in the 1970s and through the Year 2000? Part II, building upon and using the analysis from Part I, discusses these new relationships, issues, and tensions among science, scientists, policymakers, and policies. Thus Chapter 8 deals with newly emerging relationships between scientists and urban affairs, Congress, courts, economically and politically developing areas, and the states. What is scientists' emerging relationship with, and contribution to, these new areas?

Part II (Chapter 8) also discusses the contributions of scientists to many new fields, such as population control, pollution elimination, and social problems. Controversies and issues raised in such new fields are outlined—racism, manipulation of human beings, drug abuse, privacy, copyright and patent rights, and property rights in conflict with environmental rights.

Finally, Chapter 8 analyzes the relationship of the scientist to democratic and totalitarian systems. How do ideologies, religions, and political objectives distort and warp science and scientists' contributions to policy and participation in policy making? How does a democracy shape and control scientists' participation and contribution within the political process? What are the dangers and pitfalls accompanying a society's relationship with its scientists? What do scientists hold for the future, and what does the future hold for scientists?

PART ONE.
SCIENCE, SCIENTISTS, AND AMERICAN NATIONAL POLICY: THE POST-WAR YEARS, 1945-1968

Knowledge, Politics, and Power in the Executive Branch

CHAPTER 1

Bertolt Brecht's *Galileo* dramatically poses a dilemma fundamental to the scientist and his society:

> Barberini: But Bellarmin, you haven't caught onto this fellow. The scriptures don't satisfy him. Copernicus does.
> Galileo: Copernicus? "He that withholdeth corn, the people shall curse him." Book of Proverbs.
> Barberini: "A prudent man concealeth knowledge." Also Book of Proverbs.[1]

A playwright's characters representing ideas and hurling scriptural justifications may do little to resolve their personal conflicts, but as vehicles they admirably pose and sharpen issues. Brecht's *Galileo* does precisely that, and underlines an issue lurking behind this particular inquiry. Scientists, should they decide to withhold their skills and findings from society, can

NOTES TO CHAPTER 1 START ON PAGE 21

greatly influence the affairs of men and the nature of our lives. Or they can disseminate and urge their skills and findings on societies and governments and likewise change policies of governments and the course of history.

What if a scientist chooses not to withhold corn, and instead actively pursues its use by policy-makers and governments? What if he reveals his knowledge, advocates its relevance to policy, advises government officials, or becomes a policy-maker himself? As a research effort, this inquiry addresses itself to such questions. Specifically, it attempts to *identify those conditions that accompany, determine, or shape the influence of scientists and their scientific knowledge on public policies and public policy making.*

Scientists' behavior must be treated just as we would treat the behavior of any individual active in politics. Thus scientists are often best seen as political actors who, like voters, legislators, and Presidents, seek to influence public policy. Sometimes, recognizing the political importance of his actions, a scientist will withhold his expertise from policy-makers. Norbert Weiner's refusal of an Air Force request in 1946 for a reprint of a paper he had written discussing guided missile technologies serves as an example of this refusal to contribute.[2] The early-1969 one-day research stoppage by nongovernmental scientists working on defense projects provides another case where scientists refused to behave as scientists building science.

But such events are both rare and symbolic gestures. Scientists do not often refuse their services or deliberately conceal scientific findings. Still, science cannot develop, become useful, or become known without scientists. And so, it is the scientist on whom research must focus. He is the actor who has skills and knowledge to offer and urge upon policy-makers. He is the actor who may refuse his skills and knowledge to society. He is the actor who may use his reputation as a scientist to influence policy often only remotely connected with his scientific field.

Even though the science he develops can shape policy,

it is the scientist rather than science who must bear the burden. We cannot locate responsibility for man's plight or poor policy choices in science and technology. They are inert, abstract phenomena, which would not have shaped lives and events had scientists and political leaders not brought them into being and had men not decided they were relevant. Therefore, this research concentrates on men with knowledge and skill and assumes that those men who produce knowledge behave with some measure of free will and choice.

We may end by blaming society and policy-makers rather than scientists for the evils and errors that flow from science and scientific contributions. But at least we begin by considering scientists, too, as political actors who, like others, seek influence on public policy making.

THE RESEARCH: SUBJECT MATTER, ORGANIZATION, AND MAJOR ARGUMENTS

The *policy-making and policy-process approach* forms the basis for this research.[3] Such an approach enables us to ask several important questions. Do different types of scientists have different impacts on policy making? Are different types of policy more or less amenable to scientists' influence? What levels of influence accrue to specific scientific fields and scientists? In what historical period were some scientists influential over particular types of policy? What conditions accompany various levels and patterns of scientists' influence? Do these conditions vary with the type of policy at stake? Where are scientists most influential? Thus specific characteristics of a policy-shaping process will affect policy outcomes and scientists' influence over those outcomes.

This research deals with the "implications of science for policy"[4] rather than "policy toward science." *Science, transmitted through the personalities and political activities of scientists, is regarded as an input, tool, or aid useful in shaping public policy.* Thus science through scientists has increasingly served the policy-making process in a relationship that began in the nineteenth century when men found that many issues could not be resolved by reference to legal precedents or "reasoning from abstract principles." [5]

The government's science policy or "policy toward science" is the reverse side of the science-government relationship and involves the support of research that might return as a relevant public policy-making input. Science policy will be one of twenty policies studied, but it is not the focus of the research. For that, we will inquire into the role scientists and scientific considerations have in shaping science policy.

Public policies are influenced more and more by individuals and considerations with scientific and technical rather than political or legal bases. Thus this research would appear to be focusing on a significant and consequential element in the formation of public policies. Robert Wood has characterized this increased relevance of science and scientists:

> This transformation of the functional responsibility of the modern state drastically alters the position of the scientist from that of a benevolent outsider occasionally affecting the affairs of state to one of a valuable insider possessing information vital to the continued survival of the system. It subtly shifts the emphasis of the persistent political question "Can we do this?" from the consideration of legal constraints to consideration of physical constraints.[6]

The inquiry covers policy making during the *1945–1968* period on the *national level of government* in the United States. Within that national level, analysis centers on *scientists' influence on policy-making activities in the Executive Branch*. Robert Wood again provides excellent rationale for selection of the Executive Branch as a focal point, contending that:

> Decisions on military strategy, space exploration, budget

> allocations, foreign policy, medical research, and transportation facilities are in the first instance almost always formulated within the executive branch, and are frequently conclusively determined there. If a small number of key legislators can be brought in, a few prominent members of the press briefed, and an advisory committee established for prestige purposes, then the "cardinal choices" are made in executive chambers. The legislative process, the prevailing public sentiment, the articulate opinion of organized interest groups operate as constraints, reduce the number of feasible alternatives, and sometimes upset the calculations of the policy-makers. They do not, however, alter the prevailing concentration of power to propose, persuade, and establish the critical policy positions which the insiders now possess.[7]

I do not deny that some policies are effectively shaped and initiated by Congress or other groups. Indeed, some policies studied in this research were so derived, as in the cases of extraction policy (private groups, industries) and pollution policy (Congress) where significant choices were made outside the Executive Branch.

The *dependent variable* in this research is the *scientists' participation and influence over policy and policy making.* And the shape of policies in turn will depend partly on scientists' participation and influence.

However, inquiry concentrates on the *independent variables* that shape and accompany scientists' participation and varying levels of influence (dependent variable). A major factor affecting scientists' influence, it will be argued, is *the perception of significant political groups and individuals regarding the impact of scientists' participation and influence and the impact of the policy scientists are affecting.* But beyond this factor many other variables will be studied, some of which are closely related to the major factor.

Specifically, these factors that might shape or accompany scientists' influence are the visibility of policy-making, extent of support for a policy in the general political climate, sense of urgency surrounding policy making, degree of development of the policy arena, division or indecision among leadership in the Executive Branch, congeniality of the policy to private enterprise, absence of hostile vested interests, orientation of political executives, particular scientific fields involved, degree of specialization achieved in those scientific fields, scientific component of the policy or policy-making process, hypothetical character of the science involved, difficulty experienced by political officials in comprehending the science, absence of nonscientific expertise, absence of political conflict among scientists, absence of scientific conflict among scientists, competence of scientists in scientific aspects of the policy, extent of power as a value in scientists' personalities, vested interests of scientists, access available to scientists into policy making, formal policy-making positions of scientists, and functional stages through which scientists contribute to policy making.

The relationships between these independent variables and scientists' influence (dependent variable) are studied in the formation of *twenty* types of policy. Because scientists have varying levels of influence on these twenty types of policy-making processes and policies, their participation and these policy types can be grouped into *high-, moderate-, and low-influence policy processes.* Further differentiation is obtained by subdividing the high-influence category into "high" and "moderately high" and the low-influence category into "low" and "moderately low." And finally, these twenty policy types are also grouped into *nine policy arenas,* according to the policies' generally perceived impact. The resulting groupings are shown as a summary in Table 1.

Science has been defined as an input into the policy-making process. It is not itself an arena of politics or a set of political relationships and demands. However, the scientists' influence as bearers of science and participants in policy making is shaped by the general pattern of expectations that characterize various policy arenas. It is also shaped by factors stemming from the science itself and scientists themselves.

Preliminary examination of scientists' participation in

Table 1—Policy Types, Policy Processes, and Policy Arenas

Policy Process by Scientists' Level of Influence	Policy Arena by Impact	Policy Type
High	Entrepreneurial	Science
Moderately High	Entrepreneurial	Space
Moderately High	Communal Security	Weather
Moderately High	Communal Security	Weapons
Moderately High	Communal Security	Deterrence and Defense
Moderately High	Communal Security	Health
Moderately High	Economic Management	Fiscal and Monetary
Moderate	Regulative	Pollution
Moderate	Regulative	Conservation
Moderate	Regulative	Antitrust
Moderate	Regulative	Transportation Safety
Moderate	Regulative	Trade and Balance of Payments
Moderately Low	Distributive	Agriculture
Moderately Low	Distributive	Transportation
Moderately Low	Extra-National	Foreign Aid
Moderately Low	Extra-National	Disarmament and Arms Control
Low	Extra-National	Foreign Political
Low	Governmental Redistributive	Organization
Low	Social Redistributive	Social
Low	Self-Regulative	Extraction

policy making suggested that many important factors conceivably shaping or accompanying scientists' influence were related to the *political context surrounding policy formation*. Such relevant factors would include *support from the general political climate, existing patterns of political demands,* and the *perception by the Executive Branch of how politically significant groups perceive the impact of governmental activity or policies in a given field.*

Thus it seemed appropriate to use Theodore Lowi's concept of an *arena* of policy making.[8] Accordingly, scientists' influence within the Executive Branch would be significantly determined by perceptions of the impact of their activities or government's activities in a specific policy field. As the perceived impact would vary, so would the policy positions and objectives of the Executive Branch vary. In turn, scientists' influence and usefulness within the Executive Branch would be affected.

Adoption of this perspective for discussing scientists' participation enables this research to build on the foundations already laid by other scholars. Theodore Lowi's development and refinement of the policy arena as an analytical tool is basic to this research. However, the work of Robert Salisbury and John Heinz on "integrated" and "fragmented" *demand patterns* has also been used. Coupling these integrated and fragmented demand patterns with the concept of integrated and fragmented *decision systems,* Salisbury effectively built upon Lowi's work. Integrated demands and an integrated decision system lead to a *redistributive* arena. Integrated demands and a fragmented decision system may produce *self-regulation.* Fragmented demands and a fragmented decision system lead to a *distributive* arena, and fragmented demands and an integrated decision system produce *regulation.* This research considers the integration or fragmentation of a decision system (orientation of political executives, division or indecision among political leadership) and the pattern of demands from politically significant groups. Finally, James Rosenau's development of the *issue area* as a concept has considerably aided this analysis.[9]

The concepts of distributive, regulative, and redistributive arenas are derived from Lowi's original work, while that of a self-regulative arena comes from Robert Salisbury, John Heinz, and Robert Eyestone.[10] The peculiar character of scientific inputs and scientists' participation in the policy types studied, however, demanded addition of five related but new arenas. These were necessary for full understanding, given the focus on scientists as inputs rather than on policies or policy outcomes. The five are the extra-national, social redistributive, communal security, governmental redistributive, and entrepreneurial arenas.

Given the concept of an arena, it is expected that scientists' influence, because it must develop within larger political relationships, must, like those relationships, be "determined by the type of policy at stake, so that for every type of policy there is likely to be a distinctive type of political relationship."[11] In defining and categorizing policies, then, they will be treated "in terms of their impact or expected impact on society."[12]

The specific and immediate objectives of the research were twofold. First, it was an attempt to ascertain if *existing patterns of demand and politically significant groups' perceptions of scientists' impact or policy's impact would affect the influence of scientists on policy making within the executive branch.* Second, it sought to demonstrate that *factors stemming from science or the scientists would modify scientists' influence within the limits placed by groups' perceptions, demand patterns, and the policy arena.*

It will be shown that these factors stemming from science and scientists (e.g., competence of scientists, specialization, "power" as a value in scientists' personalities) may depend both on existing patterns of political demands and on the total resources that the social system will invest in scientific development.

This brief summary of the research and its basic concepts should suffice as an introduction. However, full definitions of concepts, variables, and relationships will be developed in Chapter 2.

The narrative of the research follows a simple pattern. Previous studies of scientists and policy making will be briefly discussed and conceptually related to this research (Chapter 1). The specific boundaries, relevant variables, concepts, and expected relationships among variables will be defined (Chapter 2). Broad conceptual definitions of the nine policy arenas will be set forth (Chapter 3). The general conditions that might accompany or produce scientists' influence will be identified, and the scientists/policy-making relationships will be grouped into high-influence, moderate-influence,

and low-influence policy processes (Chapter 3). These high-, moderate-, and low-influence processes will then be described in separate chapters (Chapters 4, 5, and 6) as follows:
1. Summary of scientists' participation and the policy process (high, moderate, low).
2. Discussion of the arena or arenas evidencing that level of scientists' influence.
    a. Definition of the arena.
    b. Brief characterization of scientists' participation and influence within policy types in that arena.

Finally, the narrative deals with the whole scientist/policy-making system. The research and findings from Chapters 4, 5 and 6 are summarized and immediate implications for scholars drawn (Chapter 7). This synthesis provides an opportunity to extend the analysis and consider its implications for emerging issues, new tensions, and continuing debates in the relationship of scientists to policy making (Chapter 8).

The most comprehensive objective of this empirical effort involves increased specificity about the relationships between scientists and policy making. The inquiry therefore seeks systematic identification of (1) the conditions affecting scientists' influence on policy and policy making and (2) various policy arenas which explain some of the variations and characteristics of scientist/policy-making relationships. It seeks a synthesis of previously unconnected information and observers' accounts of scientists' participation, and it should impart more organization and unity to the subject matter and permit comparability among scientists' relationships to policy making.

Data used in the research come from such secondary sources as newspaper accounts, case studies, general analyses, textbooks, and individuals' research. These sources have been merged with my own perceptions and observations.

The inquiry has importance insofar as the issues and policies discussed harbor outcomes or potential rewards and deprivations that many men would consider "significant consequences" for their lives and futures. Citation of policy types —such as weapons, fiscal, health, and agriculture policy— and possible consequences—such as continued human survival, economic sufficiency, physical well-being, and a full stomach—should make this importance more apparent.

PREVIOUS STUDIES

> In holding scientific research and discovery in respect, as we should, we must be alert to the equal and opposite danger that public policy could itself become the captive of a scientific-technological elite.

This warning, issued early in the 1960s, came from President Eisenhower in his Farewell Address. And though he later clarified his fears, limiting the danger to scientists allied with the military-industrial complex, he nevertheless raised a relevant issue for late-twentieth-century man.

But the issue and men's concern are not new. Serious efforts to cope with the subject and problem go well back in popular and scholarly literature. The volume of earlier concern never matched the outpouring we have witnessed since 1950, but the basic questions have been known and encountered since Plato.

The relationship of intellectuals, engineers, scientists, and scientific and technological elites to government and political power have been treated in four fundamental types of scholarly effort:
1. Empirical theories constructed on unstructured observations and relatively unsystematized evidence.
2. Empirical research with minimal theoretical objectives beyond its immediate focus (cf. case studies).
3. Normative and utopian theories.
4. Efforts primarily aimed at the development of concepts and frameworks for analysis.

As a summary of this basic literature, several more-important and recognized studies are listed in Table 2.

14 Knowledge, Politics, and Power in the Executive Branch

Table 2—Selected Major Works on Science, Scientists, Scientific-Technological Elites, and Political Power

EMPIRICAL THEORIES, RESEARCH, AND CONCEPTS

James Burnham, *The Managerial Revolution* (1941)
John Kenneth Galbraith, *The New Industrial State* (1967)
Waclaw Machajski (A. Wolski), *The Intellectual Worker* (1904)
Max Nomad, *Aspects of Revolt* (1959) and "Masters—Old and New," *The Making of Society*, ed. by V. F. Calverton (1937)
Don K. Price, *Government and Science* (1954)
Piet Thoenes, *The Elite in the Welfare State* (1966)
William K. Wallace, *The Passing of Politics* (1924)
Robert Gilpin, *American Scientists and Nuclear Weapons Policy* (1962)
Suzanne Keller, *Beyond the Ruling Class* ("strategic elites"), 1963
Harold Lasswell, *Analysis of Political Behavior* ("skill commonwealth"); *The Decision Process* ("functional stages" and "value transfer")
Don K. Price, *The Scientific Estate* ("establishment" and "estate"), 1965
Edward S. Flash, Jr., *Economic Advice and Presidential Leadership* (1965)
Daniel S. Greenberg, "The Myth of the Scientific Elite," *Public Interest*, I : 1 (1965)
Robert C. Wood, "Scientists and Politics: The Rise of an Apolitical Elite," *Scientists and National Policy-Making*, ed. by Robert Gilpin and Christopher Wright ("apolitical elite"), 1964

NORMATIVE AND UTOPIAN THEORIES

Francis Bacon, *The New Atlantis* (1629)
John Desmond Bernal, *The Social Function of Science* (1939)
Thomas Campanella, *The City of the Sun* (1643)
John Dewey, *The Public and Its Problems* (1927)
Aldous Huxley, *Brave New World*
Harold Lasswell, *Personality and Politics; Analysis of Political Behavior; World Revolutionary Elites* (with D. Lerner)
Karl Marx, *German Ideology; Preface; Economic and Philosophical Manuscripts*
Karl Mannheim, *Man and Society in an Age of Reconstruction* (1941)
George Orwell, *Nineteen Eighty-Four*
Plato, *The Republic*
Henri Saint-Simon, *New Atlantis; Lettres d'un Habitant de Geneve; Leaflets to Humanity*
B. F. Skinner, *Walden Two*
Jonathan Swift, *Gulliver's Travels*
Thorstein Veblen, *The Engineers and the Price System* (1934)
Lester Ward, *Applied Sociology* (1906)
H. G. Wells, *Anticipations* (1902); *The Open Conspiracy* (1928)

A sizable proportion of these efforts deals with the relationship between intellectuals, the "knowledgable," or educational elites and political power. Therefore, the usefulness of such efforts for this research depended on drawing careful analogies. Their relevance has thus been primarily illustrative, conceptual, and a contribution of logic. This use of the material has been possible despite several difficulties. These include levels of abstraction well above specific governments, policies, and historical periods and an aggregative, elite focus that obscures differences between situations and scientists. Of course, these difficulties become "difficulties" when pressed into serving my specific concerns and objectives. These abstractions, elite frameworks, concept developments, and creative, insightful observations fulfill important functions in expanding knowledge of the political system and scientists' role.

Many efforts have made significant contributions to this research, particularly in conceptualizing the relationship between scientists and the political, economic, and social sectors of societies. Therefore, it seems useful as an introduction to briefly discuss six such conceptualizations.

1. SCIENCE AND SCIENTISTS AS FUNCTIONALLY USEFUL TO SOCIETY, SOCIAL GROUPS, OR GOVERNMENTS. Suzanne Keller's *Beyond the Ruling Class*, building on Talcott Parsons' functional analysis, develops the concept of *strategic elites*. Keller's elites are powerful inasmuch as they perform or contribute to performance of the four functions set forth in Parsons' framework—goal attainment, adaptation, pattern maintenance, and integration. Scientists become socially necessary for goal attainment and adaptation functions, enabling society to control its own behavior and interact with its environment.[13]

Amitai Etzioni's *The Active Society* similarly sees a *knowledge unit* operating within a societal unit's controlling overlayer.[14] The knowledge it develops frees the societal unit from acting blindly and automatically. Furthermore, knowledge is a prerequisite to the "active" rather than "passive" orientation, as a societal unit seeks to change and control its own development. Science, for Etzioni, becomes a minor but important subsystem within the knowledge unit.

Marxist theorists share an associated pragmatic view of science's role. Marx himself saw sociology as a science providing knowledge about the real basis of society and building a scientific basis for socialism.[15] John Desmond Bernal, developing the relationship between science and Marxist practice in the 1930s, noted that science grew under capitalism but would eventually break with that system and become a "servant" of socialism. Though scientists would remain subservient to the economic and political leadership of the socialist system, they would enable "a unified and co-ordinated and, above all, conscious control of the whole of social life." Bernal's science "abolishes, or provides the possibility of abolishing, the dependence of man on the material world."[16] *The Social Function of Science,* Bernal's major work, sets forth the science-socialism synthesis within the British experience.

William K. Wallace's *The Passing of Politics* argues that economic factors have replaced political factors as the dominant force in social life and that science has become relevant since 1900 through its close association with industrial progress. Scientists thus enable the industrial sector to increase production of goods useful to men. Therein they contribute material well-being and personal happiness.[17]

Piet Thoenes' work, *The Elite in the Welfare State,* expresses a different perspective. As the welfare state develops, scientists would be increasingly used by the political leadership to provide a scientific justification for actions politically based and motivated.[18] Thoenes fears the use of science's aura of objectivity to remove issues from political contention, thereby uniting science and government in a monolith hostile to democracy and the free flow of criticism and ideas. Scientists would become the "handmaidens" of functionaries and administrators who form "an illegal elite which (no doubt with the best possible intentions) puts a halter on democracy and blindfolds science." [19]

The role of scientists and intellectuals proves not so subordinate for Harold Lasswell. He would use scholars to (1) apply psychiatric knowledge toward diagnosis of "undemocratic" personalities in leaders and toward an increase in "social health," (2) increase general awareness among the mass and set it free from distorting myths, (3) develop means through which insecurity might be reduced and the democratic values of respect and deference maximized, (4) enable foresight and anticipation of future problems and developments, and (5) develop means and techniques appropriate to elite manipulation of the masses through the use of symbols.[20]

The common element among these theorists is a view of scientists as *useful*. Scientists are functionally useful to the system or leading groups. They do not control the system.

2. SCIENTISTS AS USEFUL AND DOMINANT IN SOCIETY, SOCIAL GROUPS, OR GOVERNMENTS. The functional usefulness of science says nothing about the relative influence of scientists vis-à-vis other groups in society. Scientists might be either subordinate or dominant. But for Comte, Ward, and Veblen, they are useful and dominant.

August Comte contends that science has abdicated its proper role and become disinterested, fragmented, and specialized. However, he normatively argues that an ideal society must be constructed on material progress rather than on egoistic political liberalism or theologians' demands for "order." Therefore, Comte foresees and advocates a "social physics" that leads to decisions based on scientific calculation instead of political debate.[21] Scientists would then predominate within that society.

Lester Ward's *Applied Sociology* proposes a similar concept, with science and scientists in control of policy making. Ward's rebuttal to Herbert Spencer's Social Darwinism constitutes a rejection of laissez-faire doctrines and advocates rule by scientists. Sociologists form the basis for Ward's welfare state; they develop solutions to social problems and means to open up social bottlenecks, therein freeing men's natural affinities for one another. Legislatures, for Ward, would merely ratify and invoke solutions specified by sociologists.[22]

Thorstein Veblen's dream of an economic system run by engineers rather than by management likewise conceives of scientific and technological experts as dominant. *The Engineers and the Price System* constitutes an appeal to engineers, whose background directs them toward production and efficiency, to take control of the production process from managers, whose background directs them toward profit.[23]

3. SCIENTISTS AS USEFUL BUT CHECKED BY, OR COEXISTING WITH, POLITICAL AND ECONOMIC POWER. Don K. Price's *The Scientific Estate* abandons hope of direct popular or constitutional control of science and scientists. Price places his faith in a new set of functional checks and balances. Identifying four broad functions and functional groups in government as the scientific, professional, administrative, and political, Price relies on their competing interests and perspectives to insure checking and balancing. The functional groups are arrayed along a continuum ranging from concern and value for knowledge and truth to involvement with power and action. Scientists are an "estate" or "establishment" falling on the truth-valuing end; professionals are less truth-oriented; and the administrative and political types fall more toward the power and action end of the spectrum.[24]

Saint-Simon's work alternates between conceptions of scientists as the dominant element in a utopian society and scientists as allies of men of property desiring to avoid a class war by dispensing material progress through innovations. Indeed, scientists are often forgotten in Saint-Simon's visions, and industrialists become the dominant social force. On balance, however, he probably sees scientists and industrialists at least in alliance or coexistence.[25]

4. SCIENTISTS AND SCIENCE AS AUTONOMOUS, AND INQUIRY AS AN END IN ITSELF. Jonathan Swift's satirical account of Laputa and Lagado in *Gulliver's Travels* evidences this pers-

pective. Lagado's scientists may have had many projects in progress, most of them frivolous and silly by contemporary "standards," but none had yet come to fruition. As a result, the nation was totally impoverished by expenditures for science. The flying island of Laputa was controlled by scientists, but workmen attempting to construct scientist-designed devices could neither understand the scientists' instructions nor rely on their specifications.[26] Many observers might marvel at Swift's foresight into mid-twentieth-century science in developed nations.

5. SCIENTISTS AS A SKILL GROUP DEPENDENT ON ITS SKILLS FOR INFLUENCE AND SIGNIFICANCE. Harold Lasswell's concept of a *skill commonwealth* fits this viewpoint. His original formulation (cf. "Skill Politics and Skill Revolution" in *The Analysis of Political Behavior*) considers such skills as bargaining, propaganda and manipulation, corporate management, violence, and administration. The dominance of a skill group depends on social and historical trends, such as the demand for skills in violence described in Lasswell's "garrison state."[27]

Lasswell does not consider scientific skills specifically, but Robert C. Wood does consider them in his probing synthesis of several M.I.T. case studies:

> The limited empirical evidence available indicates that if the scientist in politics is a little less than a leader of a snow-balling political crusade, he is a good deal more than another expert performing tricks of virtuosity at the command of politicians, bureaucrats, or soldiers. He is instead a member of a small and untutored elite which has entered an inhospitable system but which possesses such valuable assets that it bids fair to displace entrenched skill groups in certain parts of the system and to establish an overall new equilibrium in the competition for power among the professions.[28]
>
> It is especially important that the influence of this group apparently does not come about by conscious adaptation to the political world, that is, by learning the skills of other elite groups, but by continuing its own sharply differentiated

behavior patterns. The group is an apolitical elite, triumphing in the political arena to the extent that it disavows political objectives and refuses to behave according to conventional political practice.[29]

This research assumes that even "apolitical" scientific activities and their impact on policy are basically "political" events. Yet Wood's insights have been invaluable in conceptualizing the scientist/policy-making relationship.

6. SCIENTISTS AS RESPONSIBLE FOR THEIR ACTIONS. Francis Bacon in *The New Atlantis* portrays scientific activity independent from the state and located in a scientific foundation. Bacon's utopian conception has members of the Academy take an oath of secrecy that prevents their communication of certain discoveries to outsiders, including political leaders and the state.[30] Thus scientists become ethically and personally responsible for their work and its results. They might well refuse to behave or act, effectively denying their skills and discoveries to societies.

Some truth lies in all these perspectives. As such, conceptualizations have power to enlighten specific aspects of a subject or view. No one way of looking at scientists and government need be more valid than another. Perspectives may be only more or less productive for specific purposes.

These authors' work provided some basic assumptions for this research. Scientists are useful to society, but only because they have unique and valuable skills. Whether scientists dominate or are dominated by society and government varies with the historical point in time and criteria used by the observer. And finally, there is a basis for assuming that scientists, as human beings and political actors, are responsible for their actions. On these basic assumptions the inquiry seeks to build a theory explaining the relationship of scientists to public policy making.

Notes

1. Bertolt Brecht, *Galileo* (New York: Grove Press, 1966), p. 77.

2. Stefan J. Dupre and Sanford A. Lakoff, *Science and the Nation: Policy and Politics* (Englewood Cliffs, New Jersey: Prentice-Hall, 1962), p. 105.

3. An *elite framework* was used in my initial explorations in this field of scientists and policy making. That effort was for a seminar on "Elites and Democratic Theory" with Dr. David Spitz at the Ohio State University. Substantially reworked, that seminar paper appears as a segment of Chapter II.

Employing the elite framework, I treated scientists collectively and policy as a single concept. I did not differentiate between types of scientists and fields of science. Nor was policy broken down into agriculture or defense policy, for instance. Such a perspective had value, however, and still poses some interesting possibilities. In distinguishing an elite as an aggregate phenomenon, one is suggesting that it makes little difference whether the scientist is a physicist, economist, or sociologist. And that it makes little difference whether the policy in question is for agriculture or defense. The *similarities* of all scientists are more important, as social and physical scientists similarly relate to policy making. The elite perspective suggests that, even though social and physical scientists may affect policy making differently, those differences are unimportant in comparison to the common influence they both exert on policy making.

But "importance" and "important consequences" are personal judgments. In suggesting in this research that the differences between social and physical scientists and between defense and social policy are "important" and "consequential" for men, I may differ with others. But because these differences are "important," we need not assume that similarities are less or unimportant. In fact, I am persuaded that the similarities are more important. But that must be a later research project. And it does not mean that the consequences of scientists' participation, as studied here, are not for me terribly important for men in American society.

4. Organization for Economic Cooperation and Development, *Science and the Policies of Governments* (Paris: OECD, 1963), p. 26.

5. Don K. Price, *Government and Science* (New York: New York University Press, 1954), p. 9.

6. Robert C. Wood, "Scientists and Politics: The Rise of an Apolitical Elite," *Scientists and National Policy-Making*, ed. by Robert Gilpin and Christopher Wright (New York: Columbia University Press, 1964), p. 54.

7. *Ibid.*, p. 66.

8. Theodore J. Lowi, "American Business, Public Policy, Case Studies and Political Theory," *World Politics*, XVI : 4 (July, 1964), pp. 677–715. An arena is a particular set of political relationships surrounding and shaping policy making.

9. Robert H. Salisbury, "The Analysis of Public Policy: A Search for Theories and Roles," *Political Science and Public Policy*, ed. by Austin Ranney (Chicago: Markham Publishing Co., 1968), pp. 151–175; Robert H. Salisbury and John P. Heinz, "A Theory of Policy Analysis and Some Preliminary Applications," (Unpublished paper, 1968 Annual Meeting of the Midwest Political Science Association, Chicago, May 2–4, 1968); James N. Rosenau, "Foreign Policy as an Issue-Area," *Domestic Sources of Foreign Policy*, ed. by Rosenau (New York: Free Press, 1967), pp. 11–50; Theodore Lowi, "American Business, Public Policy, Case Studies, and Political Theory;" and Lowi, "Making the World Safe for Democracy," *Domestic Sources of Foreign Policy*, ed. by James N. Rosenau (New York: Free Press, 1967).

10. Salisbury and Heinz, *loc. cit.*; Robert Eyestone, "The Life Cycle of American Public Policies: Agriculture and Labor Policy Since 1929," (Unpublished paper, 1968 Annual Meeting of the Midwest Political Science Association, Chicago, May 2–4, 1968).

11. Theodore J. Lowi, "American Business, Public Policy, Case Studies and Political Theory," *World Politics*, XVI : 4 (July, 1964), pp. 677–715, 688.

12. *Ibid.*

13. Suzanne Keller, *Beyond the Ruling Class: Strategic Elites in Modern Society* (New York: Random House, 1963), pp. 98–99.

14. Amitai Etzioni, *The Active Society* (New York: The Free Press, 1968), Chapters 6–9.

15. T. B. Bottomore and Maximilien Rubel (eds.), *Karl Marx: Selected Writings in Sociology and Social Philosophy* (New York: McGraw-Hill, 1964).

Notes 23

16. John Desmond Bernal, *The Social Function of Science* (New York: Macmillan Co., 1939), p. 409.

17. William K. Wallace, *The Passing of Politics* (London: George Allen and Unwin, Ltd., 1924).

18. Piet Thoenes, *The Elite in the Welfare State* (London: Faber and Faber, 1966).

19. *Ibid.*, p. 217.

20. Harold D. Lasswell, "The Political Science of Science," *American Political Science Review*, L : 4 (December, 1956), pp. 961–979; Lasswell, *The Analysis of Political Behavior* (New York: Oxford University Press, 1947). Lasswell's orientation is drawn from the whole range of his works.

21. This interpretation is drawn from Sanford A. Lakoff, "The Third Culture: Science in Social Thought," *Knowledge and Power*, ed. by Sanford A. Lakoff (New York: The Free Press, 1966), pp. 24–25.

22. Henry Steele Commager (ed.), *Lester Ward and the Welfare State* (Indianapolis: Bobbs-Merrill Co., Inc., 1967). Commager's introduction is an excellent source on Ward. Lester F. Ward, *Applied Sociology: A Treatise on the Conscious Improvement of Society by Society* (Boston: Ginn and Company, 1906), pp. 3–13, 90–113, 285–341. See also Lewis A. Coser, *Men of Ideas: A Sociologist's View* (New York: The Free Press, 1965), pp. 136–138.

23. Thorstein Veblen, *The Engineers and the Price System* (New York: Viking Press, 1934).

24. Don K. Price, *The Scientific Estate* (Cambridge: Harvard University Press, 1965).

25. Lakoff, *op. cit.*, pp. 20–21; Frank E. Manuel, *The New World of Henri Saint-Simon* (Cambridge: Harvard University Press, 1956).

26. Jonathan Swift, *Gulliver's Travels* (London: Oxford University Press, 1949); Lakoff, *op. cit.*, pp. 1–16.

27. The Lasswell works are "Skill Politics and Skill Revolution," a paper given to the Chinese Social and Political Science Association, November, 1937, and printed in *The Chinese Social and Political Science Review* (Peiping, October–December, 1937) and "The Garrison State and Specialists on Violence," from the *American Journal of Sociology*, January, 1941. Both are contained in Harold Lasswell, *The Analysis of Political Behavior*, pp. 133–156.

28. Wood, *op. cit.*, p. 44.

29. *Ibid.*

30. Francis Bacon, "New Atlantis," *Ideal Commonwealths*, ed. by Henry Morley (New York: E. P. Dutton and Co., 1885), p. 212.

The Analysis: Framework and Definitions

CHAPTER 2

Specifically, this inquiry centers on scientists developing and using science as a tool or input that may shape policy and the policy-making process. Through their own behavior, reputations as scientists, and scientific knowledge itself, these men may influence public policy.

Policies are considered for analysis insofar as they have scientific objectives, scientific components, scientific justifications, or scientific bases. And only those segments of the policy-making process within the Executive Branch of American national government are studied. This confines the research's relevance to a central government in an economically and politically developed nation. Furthermore, the limited period involved spans only the post–World War II years, 1945–1968.

Within these boundaries, the inquiry's conceptual framework and relevant variables can be identified, defined, and re-

NOTES TO CHAPTER 2 START ON PAGE 59

lated. Presumably, such precise definitions of variables could make later inclusion of more policies, time periods, nations, developmental stages, and political environments much easier.[1]

CONCEPTUAL FRAMEWORK

This book is a study of the politically relevant behavior of scientists in the formation of public policies in selected policy arenas. Beyond that some definitions are in order.

Scientists

"Scientists" will be defined broadly as individuals who are perceived to develop knowledge using scientific methods. Generally, these methods imply such norms as empiricism, control, quantification, replication, and verification. This definition should be sufficiently inclusive to cover physicists, "scientific strategists," systems analysts, economists, behavioral scientists, statisticians, engineers, and related scientific skill groups.

Engineers will be treated as "scientists" only inasmuch as their professional activities appear to be knowledge- or capability-oriented rather than application-oriented. Although the engineering sciences engage in research, testing, and a search for new techniques, engineers more often simply apply sets of rules, knowledge, and techniques in "textbook" fashion. Scientists seek new knowledge and explanations and, as in applied science and the engineering sciences, develop new technologies. Though these technologies or new bits of knowledge may be application-related or motivated by a specific prospective use, they nevertheless require and involve a development of new knowledge.

This definition of scientists rests on their particular skills and methods and distinguishes them from other skill and expertise groups such as lawyers, intellectuals, politicians, and journalists. However, these differing skills should not imply that scientists could not or will not behave like lawyers, intellectuals, politicians, and journalists. Quite the contrary, since scientists are being studied as political actors influencing public policies. If scientists merely isolated themselves in their laboratories, developed knowledge, and did nothing else, the research task would be different. But scientists *do* go beyond discovery in influencing or attempting to influence policy. And in so doing they often behave as journalists, politicians, administrators, decision-makers, intellectuals, social critics, and lawyers behave.

Science and technology

Science is a form of knowledge. Personal knowledge, moral truth, religious beliefs, philosophies, and common sense are other forms of knowledge or ways of knowing. But scientific knowledge is empirical, specific, replicable, verifiable, and sometimes quantifiable. *Technology* involves the application of knowledge as a means or "technique" for achieving a recognized, predetermined purpose. Such technologies, or standardized methods and procedures, may be derived from scientific knowledge, common-sense knowledge, or the inventor's insight. Technologies, whether they are techniques, skills, or means, often are the objectives of the applied and engineering sciences. And though they are not themselves "science," they usually rest on a new piece of knowledge or are new knowledge about accomplishing a purpose.

Technology lies close to science and scientists. As such, and as new knowledge, it can form the basis for scientists' influence. The intertwined character of science and technology is well illustrated in the fields of space exploration and nuclear weapons. Just where science leaves off and technology begins is difficult to ascertain. Though they may not produce or actually apply a "technique" or technology, scientists often develop the concept and uncover its relevance and feasibility.

Science and government are related in a myriad of

forms—science as an object of governmental largesse, source of trained manpower, instrument for international communication on a diplomatic level, sector available for pump priming and for boosting the economic growth rate, national security resource, and asset in policy making. This research deals with the last form: science as a tool shaping policy and the policy-making process.

That science-government relationship, built around the extra-scientific behavior of the scientist, may develop in several ways. Scientists may argue for the technical feasibility of a given policy; explain the nature of physical, technological or social reality; or point out the consequences of alternative policy options.

Scientists may even use their reputation as scientists to influence policy or scientific judgments when their own scientific skills are clearly not relevant. However, even this manner of influence depends on the scientific "ethos" surrounding the scientist's participation, because his influence will depend on citizens and political leaders treating his judgments as scientifically based and valid. He would still be regarded as a scientist *qua* scientist. Knowledge and the reputation for knowledge yield influence and power.[2]

Such knowledge can involve power over nature and power over people. And scientists can use science or their reputations for having scientific insight or judgment to influence policy. Scientists can offer their science to government as a basis for policy, providing indications of feasibility; outlining available means, techniques, and diagnoses; and making predictions of outcomes.

Influence

The scientists' *level of influence* is somewhat more difficult to define. Lasswell and Kaplan define influence as "affecting policies."[3] More specifically and operationally, influence is defined in this analysis as *the change in policy or the prob-*

ability of a given policy induced by the presence of the scientist and his behavior in the policy process.

Measurement of influence is obviously difficult because one must theoretically "observe" resultant policy developed without the participation of the scientist or the contribution of science. Often the entry of scientists into the policy process has changed policy, thus affording the observer a time sequence of policy with and without the key variable. However, many policy-making situations do not afford this opportunity, and the analytical task becomes more difficult. Influence, as defined, encompasses both positive (initiative) and negative (veto) effects and actions. Unintended influence will also be measured with the yardstick afforded by the definition.[4]

One problem and an important *caveat* remain. The world and policy would be vastly different without scientific discovery and innovation. Indeed, without science and scientists, modern history would have been totally altered. Had the atomic fusion process never been discovered, there would never have been a hydrogen bomb, let alone a strategy or actions based on the hydrogen bomb. The scientists' function as discoverers and therefore as scientists made all the difference.

This argument applies equally well to da Vinci and his war machines, parents who give their children knives, and deities as explanations for deadly earthquakes. Paradoxically, it is both meaningful and meaningless to blame present reality totally on scientists—and likewise to blame da Vinci for war, parents for juvenile murders, or God for natural disasters. This insight is meaningful insofar as the discoveries of scientists have made present reality possible, and meaningless insofar as so many other factors have intervened and shaped the use of their discoveries. However, the definition of influence will encompass the act of discovery of scientific knowledge, but only insofar as the discovery process is undertaken with a specific policy purpose in the scientist's mind. Discovery then becomes an explicitly political act.

Scientists thus can be treated as individuals attempting

to influence various *types of public policy*. Their behavior, considered as political activity, occurs within various *types of policy processes* and *policy arenas*.

Policy

James A. Robinson has defined *policy* in the following manner:

> Policy consists of three parts: (1) the goals, objectives, or commitments of a political unit; (2) the means selected for implementing or obtaining these goals; and (3) the consequence of the means, i.e. whether in fact the goals are actually realized.[5]

This definition is sufficiently broad to be useful for this research, since it encompasses policy *intentions*, specific policy *outputs or actions*, and eventual policy *outcomes, end results, or effects*.

The definition therefore can embrace (1) the implicit attitudes of policy-makers, (2) actual governmental behavior, and (3) such ambiguous situations as governmental inaction, non-decision making, and "laissez-faire" positions. This concept of policy may differ from stated or intended policy. Policy need not be stated or intended to be policy. Indeed, policy can be the researcher's abstraction from reality of a pattern of behavior that he then superimposes on reality for understanding. So conceived, policy is nonspecific and develops from a web of decisions and actions evolving over a period of time.

This analysis, then, focuses upon scientists' influence over the policies, policy positions, policy outputs, and policy outcomes related to the Executive Branch.[6] No direct attempt is made to study scientists' influence on the Congress, society, states, or courts. Policies are considered relevant if there is a scientific component located in the policy (monitoring stations ringing the USSR for a test ban treaty) or if the scientific method and research are used to choose or justify a most appropriate policy among policy alternatives.

Policy types

Twenty different types of policy have been selected for analysis. Policy type refers to the *substantive* content of the policy as distinguished from *target* or *institutional* categories of policy.[7] This study already focuses on an institutional category—*executive* policy and policy making.

Selection of the policies was not random. An attempt was made to include policy types harboring various different political environments, impacts, sciences and scientists, and policy-making processes. The policy types and time periods chosen for analysis are as follows:

Extraction Policy (extractive industries—oil, coal, gas, ores; fuel and raw material resource extraction), 1945–1968.

Social Policy (allocation of wealth, respect, and security to major social groups; welfare programs; ghetto development; Departments of Labor and Health, Education, and Welfare; training programs; minorities and disadvantaged groups), 1960–1968.

Organization Policy (governmental reorganization; management studies), 1945–1968.

Foreign Political Policy (normal diplomatic activity; political relationships with nations; long-range U.S. role in world; State Department), 1945–1968.

Foreign Aid Policy (allocation of A.I.D. Funds; U.S. multilateral development funding relationships; World Bank contribution), 1945–1968.

Disarmament and Arms Control Policy (Test Ban Treaty; Arms Control and Disarmament Agency; international and domestic weapons controls; nuclear non-proliferation treaty), 1945–1968.

Agriculture Policy (Agriculture Department; parity, soil bank; production research; economics of farming), 1945–1968.

Transportation Policy (Department of Transportation; mass transit support; interstate highways; Interstate Commerce Commission, Federal Aviation Agency, Civil Aeronautics Board, National Transportation Safety Board and other regulatory agencies; SST development), 1960–1968.

Transportation Safety Policy (auto safety legislation; Departments of Transportation and of Health, Education, and Welfare; regulatory agencies), 1960–1968.

Pollution Policy (air, water, noise, and waste pollution; Departments of Commerce, of Interior, and of Health, Education, and Welfare; pollution control measures), 1960–1968.

Conservation Policy (conservation of natural and recreational resources; Departments of Interior and of Agriculture), 1945–1968.

Antitrust Policy (Department of Justice; mergers; pricing; competitive factors), 1960–1968.

Fiscal and Monetary Policy (economic "health" of the nation; Council of Economic Advisers; Treasury Department; Federal Reserve System and Board; use of budget; federal spending and credit restrictions), 1960–1968.

Weather Policy (meteorology; weather modification and prediction; Weather Bureau), 1945–1968.

Health Policy (Department of Health, Education, and Welfare; National Institutes of Health; Surgeon General's Office; Food and Drug Administration; Federal Trade Commission; antismoking and tobacco legislation), 1945–1968.

Weapons Policy (weapons development; production and deployment; systems analysis; military-industrial complex; Department of Defense), 1958–1968.

Deterrence and Defense Policy (military strategy, Department of Defense; "scientific strategists"; systems analysis), 1960–1968.

Space Policy (nonmilitary uses of space; moon program; National Aeronautics and Space Administration; scientific exploration of space), 1957–1968.

Science Policy (government support of science and scientific research; National Science Foundation; National Acad-

Conceptual Framework

emy of Sciences; Office of Science and Technology; President's Science Advisory Committee), 1945–1968.

These policies are obviously not mutually exclusive. Presumably, many could be combined. However, given the focus on scientists' influence and participation, the various types do show different scientist relationships to policy making. Therefore, separation into these several types of policy is appropriate and productive.

The selected time periods all fall within the postwar years, 1945–1968. Some span that entire time. However, many policy types are confined within shorter periods (1957–1968, 1960–1968). This has been done to insure a *distinct* scientist-policy relationship, when the inclusion of earlier years might encompass a different relationship. This means we are comparing different historical and political periods, but that is precisely the point. Different historical and political conditions should differently shape the scientists' influence and participation in policy making.

Specification of exact time periods does not imply that the process outside the period was different. That probably is the case, but for periods beginning in 1945, the "cut" was made in that year merely to limit the data that must be handled. Indeed, policy types studied over the entire postwar period are likely to be stabilized arenas with stability predating 1945. Presumably, therefore, scientists' relationship to policy making has not been fundamentally altered in these fields.

Policy Process

These twenty *policy types* are gathered into types of *policy process*. This typology is based on the level of scientists' influence within the executive over the various policies. The division is tripartite—high-, moderate-, and low-influence policy processes. Wilfrid Harrison's definition of process serves well for our purposes. Harrison argues that process "seems to refer to a system for turning out a product, a special way of inducing changes that works along fixed and predetermined lines and is repeatable."[8] The processes, as we will describe them, will be composed of patterns of activity in specific situations. The process typology, however, depends on the level of scientists' influence, and the twenty policy types are parceled out on that basis.

Policy arenas

The twenty *policy types* have also been described through a series of *arenas*. Nine *policy arenas* have been substantively and conceptually derived from Theodore Lowi, Robert Eyestone, John Heinz, and Robert Salisbury.[9] Working from Lowi and Salisbury we define *policy arenas* in terms of policies' "impact or expected impact on the society." As such, policy arenas evidence particular sets of political relationships.[10] For these writers, policies are either distributive, redistributive, regulative, or self-regulative in terms of the impact of government activities or policy. Thus, as a foundation for analysis, Robert Eyestone sets forth the following definitions:

Redistributive policy: ". . . policy style indicates that governmental action toward one sector is to be considered explicitly in relation to governmental action toward other sectors." ". . . readjust the wealth, status, and power balance between sectors as wholes."

Distributive policy: ". . . describes a situation where the separate interests within a sector are treated individually by government because such treatment does not affect other sectors or other interests within the same sector."

Regulative policy: "deals with individual sectors, establishing general rules which guide governmental action toward the various identifiable interests within the same sector."

Self-regulative policy: ". . . give substantial latitude to a sector to determine the exact form of governmental programs affecting it."[11]

However, this analysis uses *nine* policy arenas. Their definitions and the policies falling within the arenas' boundaries are in Table 3. Three are borrowed (regulative, self-regulative, distributive), two are adapted (governmental redistributive, social redistributive), and four are extended (economic management, communal security, entrepreneurial, extra-national) from the Lowi-Eyestone-Salisbury-Heinz frameworks. The number of arenas has been expanded due to the peculiar character of the scientist and policy-making subject matter, but the four basic types of *impact* have been retained and still apply.

Self-regulative, governmental redistributive, social redistributive, extra-national, and distributive policy arenas fall within the low-influence type of process. Scientists' influence is moderate in the regulative policy arena. Finally, the economic management, communal security, and governmental entrepreneurial arenas involve high levels of scientist influence.

Policy, according to Lowi, is shaped in an "arena of power." The particular arena of conflict is determined by perceptions of the impacts of previous similar policies. Such expectations are crucial. Thus "expectations are determined by governmental outputs or policies." [12] Lowi's functional definitions of policies therefore depend on earlier policies or decisions of the same type and on perceptions or expectations about the impact of the proposed policy.

Defining an arena as a particular set of political relationships, Lowi then notes that these relationships are determined "by the type of policy at stake, so that for every type of policy there is likely to be a distinctive type of political relationship." [13]

This research argues that scientists' influence over policy making in the Executive Branch is shaped by (1) these perceptions of impact and policy arenas, (2) factors peculiar to the science and scientists themselves, and (3) the amount of resources committed and available within the socio-political system for scientific fields. The research will not deal directly with the third factor but assumes that *system resources* do indeed shape public policy and outcomes, at least insofar as they increase total capability for the use of science. However,

Table 3—Summary of Policy Types, Processes, and Arenas

LOW-INFLUENCE POLICY PROCESSES

Self-Regulative Policy Arena (private sector or group determines form and content of government programs or actions affecting that sector).
  Policy Type = Extraction.
Social Redistributive Policy Arena (redistribution of rights, skill, or wealth to deprived groups in order to increase their social respect and relevance; government, "haves," and "have-nots" as major actors).
  Policy Type = Social.
Governmental Redistributive Policy Arena (government's own action to internally reorganize and redistribute roles, functions, and power).
  Policy Type = Organization.
Extra-National Policy Arena (non-domestic groups as major benefactors of government actions seen as "distributive" and "redistributive").
  Policy Types = Disarmament and Arms Control, Foreign Aid, and Foreign Political (Diplomacy).
Distributive Policy Arena (government distributing benefits to groups not in competition; logrolling; no choice or zero-sum situations; one group's claims do not affect another's claims).
  Policy Types = Agriculture, Transportation.

MODERATE-INFLUENCE POLICY PROCESSES

Regulative Policy Arena (government as actor establishing rules for groups in competition with one another and with other sectors in society).
  Policy Types = Antitrust, Pollution, Transportation Safety, Trade and Balance of Payments, and Conservation.

HIGH-INFLUENCE POLICY PROCESSES

Economic Management Policy Arena (government as actor behaving in a regulative manner for the general interest; government shifts benefits among groups to promote larger interest of whole economy; orginally seen as redistributive, now regulative).
  Policy Type = Fiscal and Monetary.
Communal Security Policy Arena (government distributing collective and common benefits to entire society; benefits both tangible and intangible).
  Policy Types = Health, Weather, Defense and Deterrence, and Weapons.
Entrepreneurial Policy Arena (government produces benefits it produces itself and distributes widely; government with its own vested interest; government acting as an "entrepreneur" in the business of "manufacturing" or "producing" a product).
  Policy Types = Space, Science.

information on the resources of the society (wealth, industrialization, education) does not fully explain scientists greater influence over some policies than over others.

But patterns of political demands do indicate specific policy arenas where scientists are needed, scientists' efforts are wanted, and scientists' participation is welcomed. Political demands and attitudes thus become relevant in the various policy arenas through groups' expectations of a policy's distributive, redistributive, regulative, and self-regulative impact.

The nine *policy arenas,* three *types of process,* and *twenty policy* types are summarized and categorized in Table 3.

Five problems inherent in the research and the conceptual framework (Lowi) must be resolved or made apparent before proceeding further:

1. Measurement of either the impact or expected (perceived) impact of governmental policies is difficult. Unfortunately, the Lowi framework is conceptually powerful but operationally weak. However, given the power it has for demonstrating and explaining scientists' influence, it would be foolhardy to scrap the framework merely because precise measurement is impossible.

2. The policy arena categories may not be mutually exclusive. Some policy types may well fit into two or more arenas. Indeed, the arenas themselves may overlap. I would argue, however, that despite overlapping, the reasonably distinct characteristics of each area justify their separation. Furthermore, when two or more arenas can be considered together to further explain scientists' influence, that will be done.

But beyond this, suppose a single policy type such as agriculture policy harbors programs and activities with varied impacts (regulation, distribution, self-regulation) or involves changing impacts over several decades. The time dimension can be controlled by selecting specific periods for analysis. Regarding the various impacts which might accompany various programs or activities within a large policy field, it can be argued that (1) politically significant groups (whose perceptions, not scholars' detached observations, concern us) perceive a policy field generally to have a major impact and that (2) it is possible to consider a policy field as *predominantly* distributive, regulative, redistributive or self-regulative in character, particularly if the unit of analysis is the policy and not specific programs or activities.

3. Lowi's framework relies on expectations and perceived impacts to define policy arenas and relationships. Whose expectations? Whose perception? Whose point of view? Different groups often see policy proposals resulting in different impacts. A policy may be perceived by one group as redistributive and by a second group as distributive. One nongovernmental group may see a proposal's impact as redistributive and the Executive Branch itself consider its proposal as regulative. I do not feel this problem is as serious as first appears, primarily because the impacts of most policies are similarly perceived by relevant groups and decision-makers.

However, for this research, the relevant perceptions of impact will be these held by the Executive Branch. It must decide whether to use or heed scientists, and, therefore, its perceptions are crucial. Of course, the manner in which the Executive Branch perceives a policy's impact may well be colored by the manner in which politically relevant groups perceive the impact. This will become clear when particular policy types and arenas are discussed. Perceptions of all participants are relevant, and groups' conflicting perceptions may sometimes inject confusion; but the Executive Branch's perceptions are the most relevant. Other than scrapping the Lowi scheme, I see no other alternative to the dilemma. At least an *awareness* of possible conflicting "definitions of the situation" or expectations of impact is helpful.

4. The sample sizes are statistically small. The Ns equal twenty policy types, or three policy processes, or nine policy arenas. Within subgroups, the sample may dwindle to one case or type of policy.

However, the fact that the research deals with twenty national policy types is some consolation. The total number of possible policy types, using the typology employed, could not exceed one hundred. Viewed in this light, the apparently small sample becomes somewhat more significant. Furthermore, the objective of the research is more to suggest certain generalizations and arguments, more so, at least, than definitively to say all there is to say about scientists' influence within the Executive Branch. The effort is, in other words, more exploratory than hypothesis testing.

The twenty policy types include policies often formed under crisis situations. Crisis, as used here, is defined as a situation with significant threats to important values and with general agreement that decisions or policies in response to the threat must be developed in a limited time period.[14] Insofar as threatened values are concerned, scientists, unless they threaten the values, will be needed and influential in policy making. However, the more limited the available time for forming policy or decisions, the less likely scientists will be needed and influential. Consultation and development of scientific advice requires time, which is often not available. But even in situations demanding immediate decisions, scientists may have an impact through contingency planning and general preparedness. Scientists are most likely, however, to be influential when time, decentralization, and an "open" decision group allow room for reflection, full disagreement, and perhaps research.

5. Having not engaged in original research (interviews, new data collection), and limiting the inquiry to existing studies of policy making and incidental accounts of scientists' participation, poses a problem of validity. Clearly, therefore, intensive research into scientists' activities and role in policy making in specific areas must precede any definitive statements about their influence and participation. But this initial effort must be seen as a hypothetical sketch working sometimes impressionistically from a minimum of data.

DEFINITIONS AND DISCUSSION OF FACTORS SHAPING SCIENTISTS' INFLUENCE

Many factors might presumably affect scientists' influence on policy making. This research attempts to identify those factors or characteristics associated with the various policy arenas and levels of scientists' influence. Some factors may inhibit and others facilitate scientists' participation.

The purpose of the research, then, is to describe scientists' influence in a particular arena with reference to the presence of certain conditions or events (i.e., conditions presumably determining scientists' relationship to policy making in that arena). Thus scientists' low level of influence in a self-regulative arena would be determined by factors such as a hostile vested interest (extractive industries), a lack of need or support from executive leadership in government (laissez-faire), and a general political climate that is uninterested in backing an active governmental role in that field.

Therefore, these relevant factors should be defined and discussed. This will be accomplished by a brief definition of the factor and a discussion of the factor and its possible relationship to scientists' influence. In other words, each factor contributes to a hypothesis—that the factor, when operative, shapes scientists' influence in a given direction and that various factors associate with various levels of influence.

The various factors are summarized in Table 4 at the end of the chapter. However, more thorough treatment and discussion of the influence-related factors follows.

Scientists' level of influence has already been defined as "the change in policy or the probability of a given policy induced by the presence of the scientist and his behavior in the policy process." This level of influence has, for each policy type, been considered either high, moderately high, moderate, moderately low, or low.

The remaining variables fall into three types—exogenous, endogenous, and participational variables. *Exogenous variables operate outside the immediate scientific or scientist-related aspects of the policy process. Endogenous variables relate to the science or scientists themselves. Participational variables describe the scientists' relationship to the policy-making process* (cf. access points, formal positions, functional stages where participating).

EXOGENOUS FACTORS

The exogenous factors describe the political context of the particular policy-making situation. Many represent patterns of *demands* rather than *system resources*. Such exogenous factors as general support in the political climate for scientist- and executive-shaped policy, division or indecision among executive leadership, congeniality of policy to private enterprise, and hostile vested interests involve political demands and needs.

EXTENT OF VISIBILITY. Visibility depends on the *extent to which information on the policy or policy-making process enters into general circulation and awareness within governmental and public circles.* If folklore surrounding the postulated "scientific elite" is accepted, then invisibility and secrecy enhance scientists' opportunities to exert influence. The requirements of national security withdraw reams of scientific data from public scrutiny and, in many cases, prevent debate itself. Many argue that scientists can exert unwarranted influence behind this shroud.

Secrecy and a lack of awareness of policy-making activities split the nation and scientific community into two segments—those with access to information and those without access. The availability of data to "insiders only" may prevent any significant outside critique of established policy or policy proposals.

Invisibility surrounds such sensitive research as chemical and biological warfare development. Only full-time scientists on a project could conceivably keep up with research going on, and the insider is neither willing nor permitted to criticize.

Secrecy and unawareness dampen the flow of information and criticism within bureaucracies having operational responsibility. Max Weber notes the added influence that accrues to subordinates in bureaucracies because of their expertise, secrecy, and their superiors' total reliance on them for information and perspectives on reality.[15]

The imposition of security controls, secrecy norms, and a general lack of visibility effectively may shield the scientist, decision-maker, and public from one another. Indeed, visibility would enable better scrutiny, informed participation, and the involvement of more perspectives and forces in policy making. The problems of invisibility range from national security policy making to domestic policy formation in arenas in which political sensitivity requires that government agencies work and shape policy behind a curtain.

DEGREE OF SUPPORT IN GENERAL POLITICAL CLIMATE. Support exists when there is *a general state of approval outside the Executive Branch (Congress, public opinion) for the main policy thrust of scientists and their science, the participation of scientists in policy making, or the policy implications of scientists' proposals and participation.* The rationale behind this assumption should be self-evident. General support behind the direction and implications of scientists' efforts, whether the implications are linked to scientists or not, greatly facilitates their participation and influence. Surely, a hostile climate or lack of support would frustrate their efforts and objectives.

SENSE OF URGENCY. An urgent situation exists when, *within the context of the policy-making environment itself but not necessarily within the general political climate, there are significant pressures for formation of a policy, resolution of conflict, or elimination of uncertainty.* Though the public might also share the feeling, a sense of urgency must exist within the Executive Branch and among scientists. The features of the cold war and ghetto disturbances have evoked some sense of urgency.

Presumably, when time available for formation of a policy is short but not of crisis proportions, then scientific "insiders" may be more influential. Policy must be made, and political leaders do not have time to familiarize themselves with the scientific issues raised and the advice they are given. Longer time would allow scientific opposition to mobilize and countervailing expert opinion to be sought out.

Science is often required the most in situations with a surrounding sense of urgency. The urgency may stem from extreme threats to basic values or fears of intolerable consequences if action is not taken. Appropriate and successful action often is seen by policy-makers to hinge on good scientific advice. Thus the scientist may shape policy and exert influence.

DEGREE OF DEVELOPMENT OF THE POLICY ARENA. Underdevelopment is characterized by a policy-making process that *has not become routinized, stabilized, and generally accepted* and for which no tradition *for the use of scientific research or advice exists.* This factor could conceivably shape scientists' influence in opposite directions. Newly developing arenas, free from traditions and routine, might be more amenable to scientific advice and information. Their responsiveness and open quality would be more hospitable to scientists. However, it might be that the arena's newness would inhibit scientific influence because no channels had yet developed for influence to operate. Maturity at least insures that scientists have a place on the organization chart, even though tradition and ideology might be more likely to warp their participation and diminish their influence.

DIVISION OR INDECISION AMONG LEADERSHIP OF THE EXECUTIVE BRANCH. Division or indecision indicates *some measure of general policy conflict, disagreement, or inability to decide on a policy position by political executives who would use or solicit scientists' contributions.* Presumably, scientists' contributions might have been the fuel producing the conflict in the first place. However, behind this factor lies the presumption that scientists cannot exert influence in an executive environment fraught with political conflict (not conflict over scientific matters) or characterized by an inability to take a policy position. They would be most influential in a stable, purposive organization. Fundamental and persistent conflicts or indecision would be the most hostile environment for scientists' influence and participation.

Division or an inability to decide might possibly lead policy-makers to turn to science for solutions to problems for which no *political* solution could be found. This would depend perhaps on policy-makers' agreement to "let scientists arbitrate" and on the underlying conflict being so basic and irreconcilable that issues could not be decided politically without risking overt violence and social–religious antagonisms. Or divided policy-makers might each use its own team of scientists, one of whose position might eventually prevail. However, by division or indecision here I am concerned more with the argument that resultant chaos, weakness, and ineffectiveness in mission-fulfillment persisting over a period of years would render an agency less capable of using, and less willing to use, scientists.

CONGENIALITY TO PRIVATE ENTERPRISE. Congeniality exists when *government policy and the scientists' efforts coincide, promote, or do not interfere with the interests of private economic enterprise.* The policies of the Executive Branch may indicate a general laissez-faire tendency to do nothing, or merely enforce an industry's own guidelines. Or government and industry preferences may unite in pursuit of a similar long-range objective, even if for different reasons (cf. space industry, supersonic transport development). Policy may even promote private enterprise, perhaps through distributive allocations of financial support (cf. mass transit subsidies).

As an example of coinciding objectives, the "military-industrial-political-scientific complex" stands out. Ralph Lapp, calling the phenomenon a "weapons culture," sees a coincidence of scientists, defense industries, politicians, labor groups, local communities, and military interests.[16] Similar amalgams may be emerging around the problems of welfare and transportation and have long existed in agriculture and health. These mutually reinforcing interests may make it unnecessary for scientists to appeal for public support, since they would already have significant roles inside these webs of interdependence. Lapp calls attention to some scientists in the Pentagon who have grown up there since World War II with no socialization in the open world of free science and submerged in secret activities. This specialization has tended to produce an influential "group of scientists with pronounced military orientation." [17]

ABSENCE OF HOSTILE VESTED INTERESTS. The "vested interests" meant are nonscientific, nongovernmental, and reside in the private sector. Absence of their hostility involves the interests' *passive acceptance or active support of the main policy implications of the scientists' contributions or the participation of scientists in policy making*. Presumably, a hostile vested interest can be a potent death knell for scientists' participation and influence. Hostility is defined as active, including vocal, opposition to either the scientists or the policy implications of their involvement in the policy formation process of the Executive Branch.

ORIENTATION OF POLITICAL EXECUTIVES. The *attitude and response of political and administrative leaders in the executive branch to the scientists' arguments or research may involve agreement, opposition, unawareness, or indifference*. It is presumed that indifference and unawareness are major barriers to scientists' influence, and that indifference is probably worse from the standpoint of influence than opposition. Since this inquiry focuses on the scientists' influence within the Executive Branch, this factor becomes crucial.

Political executives and administrators should not be considered mere passive receptors of scientists' recommendations. They are not without resources and techniques for controlling and using the scientist, despite their lack of a scientific background and needed information.

Warner Schilling lists five strategies for the official confronted with *conflicting* scientific advice: make sure *scientific* elements are in conflict, side with a majority of the scientists, choose the option with the least potential harm or cost if it is wrong, choose the scientists' view and data that support his own policy preferences, or use his own sense of the scientific issues involved.[18]

Such strategies, however, may not work when scientists present a unanimous front. But time on the job for the political leader will give him some sensitivity to specific scientific fields. This requires continual association with a field, however, and generalists who must deal with many scientific fields may have severe difficulty following this strategy. When this happens, a political leader might take refuge in political criteria. Piet Thoenes fears this development, noting that the modern industrial and welfare state requires a large functionary elite using scientifically trained specialists. These scientists, for Thoenes, become the "handmaidens" of administrators who use their science to justify decisions. Power does not pass into new hands since traditional political figures control scientific activity and its use.[19] And Daniel S. Greenberg, attacking the notion of a "scientific elite," further notes that scientists were invited to Washington. Very few came seeking power.[20]

Scientists came to government because they were needed, and men who are needed are not without means of influence. Therefore, I shrink from the perspectives of Greenberg and Thoenes, preferring to view scientists as exercising measurable influence over policy.

ENDOGENOUS FACTORS

Endogenous factors describe the sciences and scientists involved with policy making. Some such factors are the degree of specialization, difficulty of comprehension, and competence of scientists. Many reflect the development of the science as a science. This development is due to *resources committed by the social system* (funding, educational level, societies' wealth) and specific groups' *demands* that support diversion of resources to scientific development (scientific pressure groups, culture-sustaining elites). The differential allocations of resources and groups' demands have obviously produced some sciences more developed than others (cf. physics with behavioral science).

There are two types of endogenous factors. *Substantive factors* relate to the sciences or policy content, while *behavioral factors* relate specifically to the scientists themselves.

Substantive Endogenous Factors

SCIENTIFIC FIELD. Fields are defined as *the academic subject classification of the science or sciences most involved with the scientists or scientific component of the policy*. Six basic fields can be singled out—applied or engineering sciences, physics, chemistry or biology, economics, systems analysis, and the social or behavioral sciences. The engineering sciences (space) and physics (nuclear energy) should have greater influence on policy making than the behavioral sciences. The reasons and conditions for this will be apparent in Chapters 5–7, but many are related to the science itself. The development of the science, competence of its scientists, its contribution to policy, and its subject matter often shape the influence of scientists in that field.

DEGREE OF SPECIALIZATION IN SCIENTIFIC FIELD. Specialization is defined as *a compartmentalization, subdivision, or narrow definition of tasks within the scientific discipline or disciplines relevant to the policy being formed*. Specialization presumably increases as a science matures and becomes more "scientific." And given a need for science, policy-makers would prefer and defer most to a developed science with more specialists than generalists. The term "specialization" *does not* refer to specialization in the policy type itself, but to the scientific field.

SCIENTIFIC COMPONENT OF POLICY OR POLICY-MAKING PROCESS. The scientific component varies with *the degree to which scientific as distinguished from nonscientific factors shape the content of a policy or choices within the policy-making process.* Presumably, the more scientific factors shaping policy in relation to nonscientific factors, the more influential scientists are likely to be. Scientists have generally become more relevant to policy making since World War II. Increasingly public policies are shaped around technological questions, research, and scientific issues. This condition emerges from the increased complexity of society and the technological revolution.[21] The mere size of the federal research and development budget at an annual $17-billion may provide a rough indication of government's involvement with science and the rising importance of science and scientific advice. Furthermore, rising expectations about the role and responsibilities of government in economic, technological, and social fields have led the government into new activities and roles demanding more scientific advice and capabilities.[22]

HYPOTHETICAL CHARACTER OF SCIENCE. Science is "hypothetical" *when scientists' research or inquiry does not relate to any existing empirical situation, expected specific future event, or particular policy intention.* Thus the sciences involved in policy planning, strategy making, contingency planning, or undirected development of new technologies could be deemed "hypothetical." Presumably scientists engaged in "hypothetical" scientific work would be more influential, if only because their work cannot be tested and validated. On the other hand, this renders the science "soft" and may lead political leaders to disregard or discount its scientists' advice. Examples of "hypothetical" scientific contributions to policy making lie particularly in the fields of military planning and support for basic research.

Don K. Price relates an anecdote about a general chiding a Defense Department civilian scientist about his lack of military experience. Responding to the general's needling, the scientist queried, "And how many nuclear wars have you fought?"[23] The hypothetical nature of planning is conducive to scientists' influence in defense policy making, where their techniques have proved more reliable than military experience in evaluating weapons systems.

The introduction of game theory, the use of computers in strategy building, and cost-effectiveness techniques have brought on a new breed of "scientific strategists"[24] who have partially bridged the gap between politics and technology. Replacing an earlier group of natural scientists who temporarily filled a need for planners in the first postwar years, scientific strategists encompass systems engineers, economists, psychologists, and mathematicians. They are men of knowledge who rely on what Kenneth Boulding calls "systems-thinking" rather than "folk-thinking."[25] Atomic weaponry moves planning one step back from post-attack mobilization problems to prewar decisions on deployment, stockpiling, and contingency planning. And "strategists'" influence is compounded by enormous lead times necessary before systems become operational.

Thus planning and choices must be made far in advance to construct weapons systems that may be developed and wax obsolete before they are either deployed or used. These enormous lead times pose a dilemma for the political leader purporting to direct technological development. His task demands that he leave a free rein to science so that constraints will not distort the product and lead to technical failure. He must rely on technologists and systems analysts for advice, giving these men ample opportunity for influence.

Basic research, like highly hypothetical planning, challenges notions of political control because it is by definition undefined and not purpose-related. Direction must be negligible. According to Don K. Price, the administrator's problem "is not to control things that he fully understands, but to control the development of new possibilities that he cannot understand."[26] Price also fears that the top political authorities of the government might find themselves in charge of a system that gave them too little flexibility and too little range of choice in difficult diplomatic or strategic situations, because the system had been based on decisions made by engineers and military officers and not fully understood by the politicians.[27] The point remains that scientists' influence may increase when they engage in basic or hypothetical science as a contribution to policy making.

DIFFICULTY IN COMPREHENSION OF SCIENCE. Comprehension difficulty involves *the inability of political leaders and citizens to understand the nature and implications of the scientific component of the policy-making process.* Policymakers feel that some sciences (cf. behavioral sciences, economics) are easier to comprehend at first encounter than others (physics, chemistry), though for many the human-oriented social sciences are more difficult to grasp. Probably the closer the science to its original discovery, the more a leader must rely on scientists to interpret its meaning for policy and policy making. Time and experience conceivably could produce "popularization" of the scientific argument and easier comprehension. However, some sciences' rate of development even outruns some scientists' efforts at comprehension.

C. P. Snow, discussing the significance of the Lindemann-Tizard episode, contends that political leadership "cannot really comprehend the nature—and the fallibility—of scientific judgement."[28] Legislators and administrators may learn the jargon but never fully grasp the scientist's work and arguments. Often they must take refuge in political criteria for making policy. Of course, as has been pointed out, a leader's continued exposure to scientific questions can increase his level of comprehension. But he is at a distinct disadvantage each time a new issue or scientific field is thrust upon him.

Leaders may learn. But we have no guarantee that this will occur or that untold scientific influence will subtly creep into policy before they learn. Complexity and rapid innovation in society could easily outrun any part-time effort at comprehension.

ABSENCE OF NON-SCIENTIFIC EXPERTISE. Presumably, if policy-makers have nonscientists with some experience or "expertise" in a given field of policy, and the policy-makers trust these laymen, then scientists' influence should be less. There is an absence of nonscientific expertise when no group of *nonscientist experts, knowledgeable laymen, or individuals with a store of highly valued experience is available to shape policy*. Presumably this is the case when leaders use political analysts, lawyers, career men, practitioners, and long-time observers for intelligence and advice. When nonscientific expertise is absent, then scientists' expertise becomes more useful.

Behavioral Endogenous Factors

ABSENCE OF SCIENTIFIC CONFLICT AMONG SCIENTISTS. Scientific *conflict exists when scientists disagree on scientific and technical aspects of the research or proposed policy*. Policy-makers must perceive the conflict, however, if the disagreement is to diminish or inhibit scientists' influence. Scientists are more likely to exercise influence when they are less in scientific disagreement with one another.

ABSENCE OF POLITICAL CONFLICT AMONG SCIENTISTS. Political conflict exists when *scientists disagree on political, strategic, or valuational implications of their research or aspects of the proposed policy*. Again, policy-makers must perceive the conflict if the disagreement is to affect scientists' influence. Presumably, if policy-makers identify political bases or value biases behind the scientists' research or arguments, scientists' influence may be harmed. The feeling by leaders that some science is a mask for political preferences or is politically motivated may also be detrimental to influence.

Conflict may involve scientists' differing views over the purpose of science, support for various fields in science, national roles in international affairs, enemy intentions, defense policies, and roles for science in politics. Conflict fur-

52   The Analysis: Framework and Definitions

ther arises among scientists in different institutions, interests, and policy-making roles.

Robert Gilpin has documented an historical alternation between conflict and unanimity within the scientific community over nuclear weapons policy since 1945.[29] Conflict regularly occurs over government research grants and project funding, pitting scientist against scientist and science against science.

Examples of political, strategic, or valuational conflict among scientists include:
1. "Insiders" within government or projects vs. "outsiders."
2. Basic vs. applied researchers.
3. University-based scientists vs. government-employed scientists.
4. Space vs. defense vs. biological scientists.
5. Pacifists and advocates of withdrawl vs. scientists stressing success in the arms race vs. proponents of a limited war capability and negotiations vs. scientists stressing deterrence and offensive strength.
6. Bureau scientists vs. scientists of another bureau.
7. Scientists fearing fallout (Pauling) vs. scientists urging continued testing (Teller) and antiballistic missiles (Wigner).
8. Scientists emphasizing arms control (Pauling) vs. proponents of finite containment (Bethe) vs. advocates of infinite deterrence (Teller).
9. Scientists favoring ABM deployment vs. scientists favoring reliance on more missiles, bigger payloads, and penetration aids.
10. High-energy physicists vs. nonphysicists.
11. Behavioral sciences vs. physical sciences.
12. Space scientists vs. marine biologists and oceanographers vs. medical scientists.

Conflict, it should be noted, primarily revolves around priorities, means, immediate courses of action, and relative weights to be attached to activities or policies. The disagreement occurs over priorities and mixes, not so much over ultimate goals or the total science pie. The argument remains. Scientists' influence is diminished when their energies are consumed in internal conflict and when policy-makers perceive scientists' conflict, especially if that conflict undermines their reputations for being *scientific*.[30]

SCIENTISTS' COMPETENCE IN SCIENTIFIC ASPECTS OF POLICY. Competence involves *the political executives' evaluation of scientists' abilities and capabilities in the scientific aspects of the policy or policy making.* Obviously, the executive's perception will significantly determine his use of scientists' contributions. And these perceptions of "competence" are closely related to such other factors as scientists' conflict and specialization.

The leader's evaluation of the scientist's competence is usually related to the scientist's immediate professional field, but often scientists are heard because they are assumed to have "foresight" and well-developed powers of analysis and reason. C. P. Snow argues that scientists possess such "foresight" and as evidence cites the Franck Committee report by the Chicago scientists on the Manhattan Project. Their insight regarding the nature of the world in the atomic age went far beyond the competence of Secretary Stimson and other nonscientists.[31]

Because of scientists' prestige and "foresight," policy-makers and citizens often defer to them on issues clearly beyond the scientists' *scientific* competence. The impact of Edward Teller's views on arms races is not necessarily diminished because of his peripheral competence in international affairs. Competence, as used here, however, describes a scientist's abilities in his immediate scientific field. Accordingly, policy-makers would see the physical sciences as more competent (i.e., scientific, precise, developed) than the social sciences.

POWER AS A VALUE IN SCIENTISTS' PERSONALITIES. The value "power" is *a desire among scientists potentially relevant to policy making to influence policy.* Some have argued that scientists only want truth and desire no power. However, a cursory inspection of their activities and desires indicates the contrary—scientists' concern over defense policy and the post-Apollo space program being two obvious examples. They often want to influence public policy, certainly with regard to and within their own scientific disciplines and the sciences.

Some scientists struggle a lifetime climbing the hierarchical ladder within their own discipline, and professional groups are riddled with politics. These scientists who actively seek power over fellow scientists have been socialized into a group and have savoured the taste of political power. They become "political" men, preoccupied with scientific affairs more than the acquisition of political power, but nevertheless interested in shaping public policies. Because of this political inclination, however, the notion of a long-haired scientist absorbed in experimentation and discovery is a dangerous stereotype and an especially risky foundation to set under a theory of scientists in the policy-making process.

Often political power becomes for the scientist a means of implementing the truth he values, perhaps bypassing the "corrupting" processes of majority rule, compromise, bargaining, and politics. In other words, scientists may value power as a means to establishing truth, and not only for personal benefit or the realization of their political preferences. Truth may be a correct course of governmental action or the opportunity to use technology a scientist has developed (cf. the "truth" of a weapons system that "works").

These are two of the problems with Don Price's system of checks and balances in *The Scientific Estate*.[32] Scientists may value power more than Price thinks, perhaps more than some politicians. Truth for many scientists (weapons scientists) may serve the interests of men of power (politicians who need weapons), and as such, truth may not always check and balance power.

SCIENTISTS' VESTED INTEREST. Scientists *stand to benefit materially from influencing policy*. When scientists are *conscious of this benefit and act on that interest, then their actions are shaped by a "vested interest."* The most significant vested interest of scientists lies in the whole realm of science policy-making. Scientists' vested interests might include higher salaries, research project funding, public or scientific prestige, or simply access to influencing subsequent policies. A vested interest, in sum, both encourages a scientist to influence policy and shapes the direction in which he wishes to influence policy.

PARTICIPATIONAL FACTORS

The third group of factors that might conceivably shape scientists' influence in policy arenas relates to the scientists' relationship to policy making (participation). They indicate scientists' points of access to the policy-making process, formal policy-making positions held, and stages where scientists are participating in policy making.

SCIENTISTS' ACCESS POINTS. Scientists have access when *officials are willing to hear, seriously weigh, and on at least some occasions shape policy according to arguments or demands by scientists. Scientists holding formal positions in the Executive Branch already have access.* Six basic access points for influencing the policy-making process may be identified: the Presidency, various levels of the Executive Branch, nongovernmental research organizations, the public media, Congress, and courts.[33] Presumably, some access points will be more conducive to scientists' influence than others. Generally, however, the more access the better.

Increasingly, scientists' access has been institutionalized through such positions as bureau advisers and cabinet-level science officers and through the President's Science Advisory Committee, the President's Science Advisor, and the White House Office of Science and Technology.

SCIENTISTS' FORMAL POSITIONS. Formal organization status, it has been argued, contributes to access and thus influence. However, some formal positions of scientists lead to

more influence than others. Five basic policy-relevant, formal positions for scientists are:
1. Internal decision-maker or *policy-maker*.
2. *Researcher* in government run or dominated laboratory.
3. *Adviser* to executive agency or official.
4. *Researcher* in private laboratory.
5. *Administrator* in executive branch.

The distinction between administrators and policy-makers is one of degree. Administrators tend to apply policy to cases rather than shape general guidelines. Some formal positions involve scientists in a bureaucratic capacity that calls for a bureaucratic orientation (operations-related, coordinative skills) rather than a professional orientation (abstract and conceptualizing skills).[34] It is expected that the more bureaucratic and policy-oriented positions will yield the greater influence for the scientist.[35]

FUNCTIONAL STAGE OF SCIENTISTS' PARTICIPATION. A modification of Harold Lasswell's seven "categories of functional analysis" is useful to describe scientists' activity.[36] They may perform any of the following defined functions and, in so doing, influence policy:
1. Intelligence, Discovery or Development of Feasible Solutions, Research.
2. Situation or Problem Definition, Identification of Conditions and Trends.
3. Recommendations (Official Promotion of Policy Alternatives or Goals).
4. Prescription (Authoritative Choice Making, Enactment or Specification of General Rules).
5. Invocation (Provisional Specification and Characterization of Conduct According to Prescription) and Application (Final Specification of Conduct).
6. Appraisal (Assessment of Success or Failure of Policy) and Termination (Ending Prescriptions and Policy Arrangements).

Lasswell's invocation and application stages have been compressed into one category as have the appraisal and termination stages. A new category for "situation and problem definition" has been added.[37]

Science or scientists can perform all these functions, but as Robert Merton points out about intellectuals, the earlier the scientist functions in the policy process, the more influence we would expect him to have. In the early stages, he can concern himself more with objectives than instrumentalities and operates under less constraint and visibility.[38] Indeed, if scientists successfully dominate the initial phases of policy formation, the remainder of the process may become inconsequential and a mere "playing out the hand." Indeed, discovery itself can have a significant effect on later stages of the policy-making process. And discovery is in an initial phase.

These twenty-two factors that might shape or accompany scientists' influence on policy or policy making are listed in Table 4. Their definitions complete the framework for the inquiry. The task of analysis can go forward.

Table 4—Summary of Factors Potentially Related to Scientists' Influence

EXOGENOUS FACTORS

1. Level of Visibility.
2. Level of Support in General Political Climate
3. Sense of Urgency
4. Degree of Development of the Policy Arena
5. Division or Indecision among Leadership of the Executive Branch
6. Congeniality to Private Enterprise
7. Absence of Hostile Vested Interests
8. Orientation of Political Executives

ENDOGENOUS FACTORS

Substantive Types

9. Scientific Field or Fields
10. Degree of Specialization in Scientific Field
11. Scientific Component of Policy or Policy-Making Process
12. Hypothetical Character of Science
13. Difficulty in Comprehension of Science
14. Absence of Non-Scientific Expertise

Behavioral Types

15. Absence of Scientific Conflict among Scientists
16. Absence of Political Conflict among Scientists
17. Scientists' Competence in Scientific Aspects of Policy
18. Power as a Value in Scientists' Personalities
19. Scientists' Vested Interest

PARTICIPATIONAL FACTORS

20. Scientists' Access Points
21. Scientists' Formal Positions
22. Functional Stage of Scientists' Participation

Notes

1. The field of science and public policy making needs efforts toward comparability.

2. The relationship between research and policy has received general treatment in Gunnar Myrdal, "The Relation Between Social Theory and Social Policy," *British Journal of Sociology*, 4 (1953), pp. 210-242; and Max F. Millikan, "Inquiry and Policy: The Relation of Knowledge to Action," *The Human Meaning of the Social Sciences*, ed. by Daniel Lerner (New York: World Publishing Company, 1967), pp. 158-180.

3. Harold D. Lasswell and Abraham Kaplan, *Power and Society* (New Haven: Yale University Press, 1950), p. 71.

4. I am aware of the problems in measuring influence. Such problems include anticipated reactions, symbolic learning, socialized behavior, sleeper effects or delayed consequences, and the proper order of exerted influence (i.e., who is influencing whom). James G. March has covered these problems rather well in "An Introduction to the Theory and Measurement of Influence," *American Political Science Review*, XLIX : 2 (June, 1955), pp. 431-451. However, the magnitude of the problems and the various weaknesses of measures that can overcome them dictate that the researcher at least be aware of these factors' possible effects when attempting to measure "influence." The measure of influence adopted here is derived from Robert A. Dahl, "The Concept of Power," *Behavioral Science*, II : 3 (July, 1957), pp. 201-215.

5. James A. Robinson, "The Major Problems of Political Science," *Politics and Public Affairs*, ed. by L. K. Caldwell (Bloomington, Indiana: Institute of Training for Public Service, Department of Government, 1962), p. 169.

6. I am aware that, in a larger sense, policy is made by an interaction and concurrence among various institutions, including the Executive Branch. However, because the executive is the

60 The Analysis: Framework and Definitions

major actor in formulating most policies, it seems reasonable to talk about executive "policy" and "policy making" with meaningful results.

7. The target, institutional, and substantive categories for policies come from Lewis A. Froman, Jr., "The Categorization of Policy Contents," *Political Science and Public Policy*, ed. by Austin Ranney (Chicago: Markham Publishing Co., 1968), pp. 41–52.

8. Wilfred Harrison, "Political Processes," *Political Studies*, 6 (October, 1958), p. 248. See also Vernon Van Dyke, "Process and Policy as Focal Concepts in Political Research," *Political Science and Public Policy*, ed. by Austin Ranney (Chicago: Markham Publishing Co., 1968), pp. 23–39.

9. Theodore Lowi, "American Business and Public Policy: Case Studies in Political Theory," *World Politics*, XVI : 4 (July, 1964), pp. 677–715; Robert Eyestone, "The Life Cycle of American Public Policies: Agriculture and Labor Policy Since 1929" (Unpublished paper, 1968 Annual Meeting of the Midwest Political Science Association, Chicago, May 2–4, 1968); Robert H. Salisbury and John Heinz, "A Theory of Policy Analysis and Some Preliminary Applications," *Policy Analysis in Political Science*, ed. by Ira Sharkansky (Chicago: Markham Publishing Co., 1970), pp. 39–60; Robert H. Salisbury, "The Analysis of Public Policy: A Search for Theories and Roles," *Political Science and Public Policy*, ed. by Austin Ranney (Chicago: Markham Publishing Co., 1968).

10. Lowi, *op. cit.*, p. 688.

11. Eyestone, *op. cit.*, pp. 1–2.

12. Lowi, *loc. cit.*

13. *Ibid.*

14. James A. Robinson, "Crisis," *International Encyclopedia of the Social Sciences* (New York: Macmillan and Free Press, 1968), Vol. 3, pp. 510–513.

15. Max Weber, "Wirtschaft and Geselschaft," *From Max Weber: Essays In Sociology*, ed. by H. H. Gerth and C. Wright Mills (London: Oxford University Press, 1958), pp. 233–239.

16. Ralph E. Lapp, *The New Priesthood* (New York: Harper and Row, 1962), and *The Weapons Culture* (New York: W. W. Norton and Co., 1968).

17. Lapp, *The New Priesthood*, p. 110.

18. Warner R. Schilling, "Scientists, Foreign Policy, and Politics," *Scientists and National Policy-Making*, ed. by Robert Gilpin and Christopher Wright (New York: Columbia University Press, 1964), pp. 149–152.

19. Piet Thoenes, *The Elite in the Welfare State* (London: Faber and Faber, 1966), pp. 129, 197–214, 217.

20. Daniel S. Greenberg, "The Myth of the Scientific Elite," *The Public Interest*, I : 1 (Fall, 1965), pp. 51–62.

21. Robert K. Merton has shown a close relationship between the development of science and its individual fields and the socio-economic environment (Puritanism, receptive society, political hospitality, dominance of utilitarian norms). "Science, Technology, and Society in Seventeenth Century England," *Osiris*, 4 (1938), pp. 360–362.

22. Robert C. Wood, "Scientists and Politics: The Rise of an Apolitical Elite," *Scientists and National Policy-Making*, ed. by Robert Gilpin and Christopher Wright (New York: Columbia University Press, 1964), pp. 52–69.

23. Don K. Price, *The Scientific Estate* (Cambridge: Harvard University Press, 1965), p. 132.

24. Bernard Brodie, "The Scientific Strategists," *Scientists and National Policy-Making*, ed. by Robert Gilpin and Christopher Wright (New York: Columbia University Press, 1964), p. 240.

25. Consult Joseph Kraft's discussion of "folk" and "systems" thought in *Profiles in Power: A Washington Insight* (New York: New American Library, 1966), pp. xxi–xxiii.

26. Price, *op. cit.*

27. *Ibid.*, p. 151.

28. C. P. Snow, *Science and Government* (New York: New American Library, 1962), p. 82. This volume contains Snow's Godkin Lectures at Harvard in 1960.

29. Robert Gilpin, *American Scientists and Nuclear Weapons Policy* (Princeton: Princeton University Press, 1962).

30. Political preferences and values of scientists often find their way into the strategies, systems, and problems areas they define, develop, and research. These nontechnical, nonscientific premises intrude through (1) the emphasis and choice of problems for investigation, (2) assessment of the relevance of various facts, and (3) the implications drawn from the research. Choices of problems for research might also be influenced by the availability of public funds for a given research project, but this does not directly relate to the scientist's influence. These problems as well as the general problems of bringing science into use in policy making are discussed in Charles Y. Glock, "Applied Social Research: Some Conditions Affecting Its Utilization," *Case Studies in Bringing Behavioral Science into Use*, ed. by Charles Y. Glock et al (Stanford: Institute for Communication Research, Stanford University, 1961), Vol. I, pp. 7–11, 16–19.

31. Snow, *op cit.*, pp. 72–73.

62 The Analysis: Framework and Definitions

32. Price, *op. cit.*, pp. 118–9, 135, 153, 190–4.

33. The scientists' use of the media and molding of public opinion should not be underestimated. Indeed, the public sees scientists as "nonpolitical" and outside the web of government. They are held outside bargaining and political conflict and represent disinterested, rational argument. Scientists, Robert Wood points out, inherit the myth and symbol of the Age of Reason. All of this is true, and yet scientists' names are never recognized by the public. Wood, *op. cit.*, pp. 60–70.

34. Scientists may have a hostility to the secrecy, hierarchical structure, and discipline of a bureaucracy, but we would expect that their influence over policy would be greatest there. Lewis C. Mainzer contrasts the scientist's utopian creed with the administrator's realistic creed in "The Scientist as Public Administrator," *Western Political Quarterly*, XVI : 4 (December, 1963), pp. 814–29.

35. Some have argued that the presence of scientists in *formal* positions increases both public and political control over their behavior and influence. That may be true, but formal access does either reflect and/or increase influence. Even though scientists are formally responsible in these positions, they remained skilled voluntary participants and can bargain for their objectives using their skill and volunteer status. They face nowhere near the command or superior-subordinate relationships most political officials must face. Cf. Wood, *op. cit.*, p. 56.

36. Harold D. Lasswell, *The Decision Process: Seven Categories of Functional Analysis* (College Park, Maryland: Bureau of Governmental Research, College of Business and Public Administration, University of Maryland, 1956), p. 2.

37. These categories for functions performed by science or scientists are not the only feasible set. Anthony Downs' excellent list of functions for scientific advice proceeds from a different, but useful, standpoint. Downs, viewing an economist's function from the standpoint of the user or politician, finds nine functions: (1) providing aid in making the right decision, (2) settling internal disputes objectively by agreement to scientific arbitration, (3) investigating prior decisions made by an interested predecessor, (4) supplying "justification" for decisions made on political grounds, (5) developing rationale for expanding pet programs or proving their worth, (6) disproving a rival organization's wisdom in expanding, (7) raising new issues or revealing a need for new programs, (8) threatening to study a competitor and weaken his base of support, (9) deferring immediate demands for action and using research to buy time or "study the problem to death." Downs, "Some Thoughts on Giving People Economic Advice," *American Behavioral Scientist*, IX : 1 (September, 1965), pp. 30–32.

38. Robert K. Merton, "The Role of the Intellectual in Public Bureaucracy," *Social Theory and Social Structure* (Glencoe, Illinois: Free Press, 1949), pp. 169–171.

Scientists in the Executive Branch: Influence and Policy Arenas

CHAPTER 3

Scientists' influence within the Executive Branch on the formation of the selected twenty policies has been described as high, moderately high, moderate, moderately low, or low. Each type of policy, according to politically significant groups' perceptions of its impact, has been categorized in an appropriate policy arena. Then each policy type (e.g., antitrust, pollution, etc.) has been described with reference to scientists' level of influence and the appropriate policy arena. Graphically, the results appear in Table 5.

No numerically logical "plot" will develop when one axis is a set of nominal categories (policy arenas) and the other is rank-ordered (level of influence). Thus, these arena categories could have been arranged in any fashion. However, when the arenas were arranged as they now appear, a distinct and interesting relationship

NOTES TO CHAPTER 3 START ON PAGE 71

Table 5—Policy Types by Policy Arena and Scientists' Level of Influence in the Executive Branch

Policy Arenas

Scientists' Level of Influence	Self-Regulative	Social Redistributive	Governmental Redistributive	Extra-National	Distributive	Regulative	Economic Management	Communal Security	Entrepreneurial
HIGH									Science
MODERATELY HIGH								Weather, Weapons, Deterrence and Defense, Health	Space
MODERATE						Pollution, Trade and Balance of Payments, Conservation, Antitrust, Transportation Safety	Fiscal and Monetary		
MODERATELY LOW				Foreign Aid, Disarmament and Arms Control	Agriculture, Transportation				
LOW	Extraction	Social	Organization	Foreign Political					

occurs. Some policy types and arenas fell into the lower-influence regions, some in the moderate-influence range, and some in the high levels of scientists' influence.

The analysis offers an explanation of these relationships between policy type, policy arena, and scientists' influence. There are other explanations for scientists' influence on policy making, but this approach seems to tie together a wide range of relevant factors, such as scientists' competence, political demands, scientists' desire for power, and the nature of the scientific subject matter.

Describing scientists' influence is a matter of judgment. No quantitative measure or visible evidence exists to ease or routinize the task. Still, description however imprecise is not impossible. And judgment can be aided with the specified definition of "influence" already set forth. If we mean X by "influence" and Y by "policy," then that much more consistent will judgments be. Therefore, I trust that my analyses of scientists' participation in the twenty policy-making situations (Chapters 4, 5, and 6) will justify my judgment of scientists' influence on policy making in the Executive Branch.

The judgments or descriptions of scientists' influence and the policy formation processes are the work of one observer working with case studies, newspaper accounts, and other empirical studies. The accounts remain somewhat impressionistic and broadly sketched.[1]

The twenty policy types were described in relation to one another, rather than in relation to any outside standard or other policy types. This causes problems for comparability if additional policy types were introduced to the analysis. However, the problem does not turn out to be as threatening in practice, since "high" influence ratings have tended to be both absolutely "high" and relatively "higher" compared to the lower ratings.

Scientists' level of influence, it must be remembered, has been studied *only within the Executive Branch or executive segment of the policy-formation process.* For purposes of simplification and analysis, influence has been compressed from five to three categories. Policy types with low or moderately low scientist influence have been combined into a *low level of influence process*. Likewise, moderately high and high policy types have been compressed into a *high level of influence process*. The moderate level of influence policy types remain as they are, comprising a *moderate level of influence process*. This produces three basic types of policy-making processes, evidencing low (Chapter 4), moderate (Chapter 5), and high (Chapter 6) levels of influence for scientists. And each process has its own set of policy types and policy arenas.

The twenty policies and their formation will be described by their general characteristics and the factors associated with scientists' influence and participation. Influence-related factors, such as level of visibility and scientists' competence, will be characterized generally as high, moderate, or low, just as "influence" is described. Influence-related factors, such as scientists' formal positions and scientific fields, will be described through whatever categories are relevant to their participation, such as policy-maker, adviser, or researcher and physicist, economist, or behavioral scientist.

These characteristics and factors associated with one policy type or one policy arena can thus be compared with the descriptions of the other policies and arenas. And a policy arena (e.g., communal security), comprising many policy types, can be characterized according to scientists' influence and the conditions associated with their participation in policy making. Furthermore, various factors can be studied with reference to scientists' influence. This, hopefully, will determine how strongly each factor contributes or relates to influence.

Overview and Summary of the Analysis

Policies have been grouped into various levels of scientists' influence and characterized through factors associated with scientists' participation in policy making in each field.

Scientists in the Executive Branch 67

These policy types and levels of scientists' influence have been placed in appropriate policy arenas, according to the perceptions of policies' impacts held by political leaders and politically significant groups.

Several factors associated rather significantly with scientists' influence stem from the political sphere. Support from the general political climate seems to be helpful to scientists' participation. Likewise, influence would appear to be stimulated by support from political leaders and political executives. These *patterns of demand or support* may also affect scientists' competence, the need for scientific components in policies, and the difficulty experienced in comprehending some sciences. Political demands and supports produce an *allocation of the economic system's resources* favorable to some sciences, which then develop more competence and may become less comprehensible to policy-makers. Furthermore, factors inherent in scientists' *patterns of participation* (formal positions, stages of activity, access points, "power" as a personality component) are political in nature.

Thus a conceptual framework constructed around the concept of a "policy arena" seemed appropriate. Such "arenas" are defined in terms of demands, perceptions, and expectations related to policies and their potential and actual impacts. The analysis contends that these patterns of demands and perceptions of policies' impacts are crucial to scientists' influence. Such patterns and perceptions give influence its unique character in the different arenas and policies.

The analysis in Chapters 4, 5, and 6 will be organized as follows:

1. General discussion of policy processes with high, moderate, or low levels of influence.
2. Definition of a specific policy arena within the policy process.
3. Description of common elements in policy types within the specific policy arena.

4. Description and discussion of specific policy type and process of its formation.

Chapter 4 will treat low-influence policies and arenas. Chapter 5 discusses those with moderate levels of scientists' influence. And Chapter 6 concludes with an analysis of policies with high levels of scientists' influence.

Before becoming so immersed in specific arenas and policies, it would be best to summarize the general arguments or theses running through the analysis. Hopefully, this will ameliorate any tendency to become mired in the "trees" and unable to view the "forest." The summary involves three major points:

1. Scientists' activities in discovering, inventing, or developing new knowledge or technologies are essential *prerequisites* to influence, but they do not in themselves augment influence unless political leaders use scientists' contributions as they see or find them.

2. Scientists' access, formal policy-making positions, and activities in key functional stages of policy making are conducive to influence. But such factors related to their participation are dependent on the resources and support they have been allocated by the political and economic systems. Such resources and support produce more "competence," scientists, and positions within government. Status within the policy-making process is also conducive to influence, particularly for scientists whose personalities show an orientation toward "power" as a value.

3. Scientists' influence depends to an extent on the perceptions politically significant groups hold about the impacts of policies and scientists' participation in their formation. Building on this finding, the analysis then shows that scientists will be more influential within the executive branch when:

    a. The policy-making process is nonzero sum (distributive, communal security, and entrepreneurial arenas) rather than zero sum (redistributive, self-regulative

arenas), with some explicable exceptions (regulative, extra-national arenas).
b. The political climate and executive leadership need or welcome scientists' participation, contributions, and influence on policy and policy making.
c. No hostile vested interest is aroused or offended by scientists' participation or influence.
d. Scientists' own vested interests or values are threatened or potentially rewarded.
e. The Executive Branch must meet popular demands for security or beneficial programs that it cannot afford to neglect or fail to meet (errors being fatal at the polls and in terms of national survival).
f. The Executive Branch, acting in its own or the public's interest, becomes an actor or combatant in the society or political process and seeks to deprive or choose among sectors in society or courses of common action (e.g., regulative arena when scientists can offer rationale, means, and strategies for the Executive Branch's actions).

Conversely, scientists will be less influential or uninfluential when:

g. The political "game" is perceived as zero sum in its outcome or impact (redistributive arenas) and the Executive Branch is unwilling or unable to enter as a "third person" that would redistribute values or change payoffs in the game.
h. Changes in the existing organization of the Executive Branch itself suggest or threaten an existing distribution of authority, power, and status (governmental redistributive arena).
i. The impact of the policy or scientists' participation is procedural rather than substantive and may change the rules of the "game," legitimacy of theretofore excluded interests, or basic manner in which policy is made (self-regulative, governmental redistributive arenas).

70 Scientists in the Executive Branch

j. The Executive Branch, for political or motivational reasons, decides it will engage only in symbolic rather than tangible actions in a specific policy field.

These are the basic threads running through the analysis. Chapter 7 makes them more explicit and relates them to the various policy arenas and policies, which are individually analyzed in Chapters 4, 5, and 6.

Notes

1. Had time and the preliminary nature of the research so indicated, a *panel approach* might have been used. Under the panel technique, recognized observers, participants, or students in each type of policy making would have been randomly chosen, formally consulted, and systematically questioned. This technique is an excellent tool and should be used in subsequent work of this type.

I would only point out, however, that these observers have a direct impact on this research. Their studies have been consulted and used as basic data. This does not mean that a "panel" approach would not lend systematic, added verification for my arguments. It does mean, though, that my use of these observers' published works has produced a quasi-panel situation. The judgment of various experts has been indirectly polled, though actual questioning of a selected group (panel) was not used.

Still, this work must build on fragmentary data. Some excellent accounts of scientists' participation are available, but only in limited areas of policy making. And much other data must be extracted from general studies of policy making or restricted case studies. Thus, in a sense, theory, built on bits-and-pieces of data, precedes systematic data collection. But as such, a theory or explanation can guide further collections of data and design of studies.

The Policy-making Process and Low Levels of Scientist Influence

CHAPTER 4

Scientists' influence has been lowest in the processes forming *extraction, social, organization, foreign aid, foreign political, disarmament and arms control, agriculture,* and *transportation policies.* Extraction policy involves a *self-regulative arena.* Social policy and organization policy are shaped in *social redistributive* and *governmental redistributive arenas* respectively. Foreign aid, foreign political, and disarmament arms-control policies may be seen as *extranational arenas.* Finally, agriculture and transportation policies are basically *distributive arenas.*

Varied perceptions of impacts can shape a policy process so that scientists have a low level of influence. The impact may be a threatened shift in government's role in the policy-making process (self-regulative arena), redistribution of either

NOTES TO CHAPTER 4 START ON PAGE 127

status (social redistributive arena) or power within the Executive Branch (governmental redistributive arena), distribution or redistribution of benefits to foreign countries (extranational arena), or government's distribution of benefits to groups in the private sector (distributive arena). Scientists' participation is seen both by the politically relevant and the Executive Branch to contribute toward each of these policy impacts.

And since some impacts are politically unacceptable, the executive shies away from policy in those arenas, neither wanting nor needing scientists' influence or participation. Redistribution, when perceived as the impact of a policy, diminishes support from groups that *have* whatever value is to be redistributed. In this manner redistribution threatens such values as the power to shape policies affecting oneself (self-regulative arena); respect, rectitude, affection, or well-being (social redistributive arena); power to shape policy and others' behavior (governmental redistribution arena); and possibly the redistribution of wealth, respect, skill or well-being from the domestic "haves" to foreign countries (extranational arena). Finally, when a policy is seen to distribute values to private groups in the society, general support accrues to the policy. The Executive Branch can then merely dispense benefits without having much competence or scientific capability of its own (distributive arena).

Scientists in these arenas are relegated to formal status *outside* government. They enter primarily through the lower levels of the bureaucracy. Congressional support, except for the logrolling politics of distribution, is absent. Scientists have developed no existing vested interest to pursue, though influence might lead to eventual positions and research funds. Still, they show no meaningful power orientation or desire to shape policy. Levels of scientist specialization are low.

Policy-makers are less likely to see some of the sciences involved as "scientific" or competent, particularly the social and behavioral sciences. These social and behavioral sciences may have a significant role in evaluating policies and shaping opinion through the educational process, but their overall impact is not very great. Scientific considerations only minimally shape policy in these arenas, despite that some of the sciences involved are relatively easy for laymen to comprehend.

The general political climate lacks support for governmental activity in some low-influence arenas, especially when the impact is "redistributive." The political leadership of the Executive Branch is often indifferent or somewhat opposed to scientists' participation and the implications of their participation. Often hostile vested interests such as the government, State Department, and oil industry oppose change or government activity, and no significant congeniality to private enterprise is involved.

No high sense of urgency for change or governmental activity exists. Visibility and political conflict among scientists and political leaders are invariably the rule. The arenas are a vortex of politics and leaders are reluctant to act in the face of risks in policy making. Distributive policies are an exception, logrolling being a well-established and supported mechanism.

Summarily, the *executive branch either does not need scientists (distributive policy arenas) or does not want scientists (redistributive policy arenas) because the impact of their participation is "redistributive," a direction the political executive is reluctant to pursue, or "distributive," an impact the government does not need scientists of its own to achieve.*

THE SELF-REGULATIVE ARENA

The fundamental political questions in a self-regulative arena revolve around "structural" issues rather than "allocative" issues.[1] Such structural rules governing the policy-making game are basic issues and may determine who will prevail. Therefore, present allocations will not be as important as the structures or rules guiding future allocations.

The major issue in self-regulation is the role of government or the Executive Branch in policy making. Self-regulation means that government engages in minimal activity, perhaps just enforcing the policies made by the private sector or lesser governmental levels (cf. Interstate Oil Compact Commission, Texas Railroad Commission).

The threatened impact of any change would involve *redistribution* and an increase in government's role as an actor, perhaps regulating the previously autonomous private sector. The new rules would presumably mean that government would do more than react to, support, and enforce private initiatives. "Laissez-faire" arrangements would no longer prevail. Private interests, such as the oil industry, would have been unable to fend off government regulation or intervention. Regulated interests might find they no longer controlled the regulatory agencies they had long ago "captured" and used to distribute public benefits to themselves or enforce industry agreements.

But many private interests are well-established and were firmly entrenched before government itself became strong. The oil industry, for example, grew up when government was no threat as a potential entrepreneur or competitor. And when a threat emerged that government might itself develop oil shale lands it owned, the industry was sufficiently entrenched to inhibit government activity. Such was not the case with atomic energy and space development, and differing structural arrangements emerged between public and private sectors.

The "game" is perceived as *zero-sum*.[2] Private economic groups dominate the process shaping their own policy. They do so with Congressional support coupled with either the acquiescence or impotence of an executive agency. The agency therefore is (1) prohibited from using scientists or developing a scientific capability, (2) reluctant to use scientists, or (3) does not want to use scientists. Scientists would increase the agencies' capacity for autonomous action, either to move from the role of "enforcer" to "regulator" or to extract resources for its own benefit (cf. oil shale). The private economic groups would prefer to monopolize all scientific development in their field. Therefore, these industries' integrated demands, as Salisbury would argue, have prevailed over a fragmented or weak decision system. As a result, I would argue, scientists have been excluded and uninfluential.[3] Government in these situations neither needs, wants, nor uses science.

Often self-regulation appears formally as regulation. However, as Murray Edelman points out in *The Symbolic Uses of Politics,* much "regulation" is only "symbolic" rather than "tangible."[4] Regulation is not really regulation in such cases and can be better understood as self-regulation. Government "regulation" of extraction industries is an example. The earlier era of transportation "regulation," when interests captured control of independent regulatory commissions, was a similar situation. Population policy, now left on a *laissez-faire* basis to families' self-regulation, would threaten many should government move toward tangible regulation and beyond "symbolic" concern with population growth.

The Executive Branch leadership knows that potential scientific contributions exist but is indifferent to their use in shaping policy. Policy, which means support of the status quo, is seen to have a minimal scientific component. Therefore, the economic and applied sciences most relevant to the self-regulative arena are denied formal policy-making positions. These sciences have minimal access to government—and this only at lower levels of the bureaucracy. Scientists neither prescribe nor invoke policy, nor do they effectively advise policy-makers.

Scientists' influence is *low* also because their participation is opposed by a hostile vested interest and because their activities in government are likely to be harmful to the perceived interests of private enterprise. The scientists involved have no significant power orientation or vested interest. They do not evidence any overwhelming desire to shape policy, though many would prefer to see it changed. They have no significant scientific conflicts among themselves. No sense of urgency surrounds or inhabits the self-regulative policy arena that might stress change, demand inclusion of scientists in policy making, or encourage a more active governmental role.

In sum, the science and scientists are available and progressing, but they are not yet meaningfully involved in policy making within the Executive Branch. They remain either in the university or private sectors, independently working away or serving the interests of the self-regulated groups.

Scientists in a Self-Regulative Arena: Extraction Policy

Scientists working or attempting to penetrate within the extraction policy process face a solid and forbidding front in the private enterprise sectors of oil, coal, minerals, and other extraction industries. The relevant scientists, largely statisticians, economists and extraction engineers, have minimal influence over policy. Scientific considerations have very little to do with existing policy or government's role in this self-regulative arena.

There are, indeed, scientists (economists, statisticians, resource engineers) employed by the extraction industries, but they are involved in policy formation only insofar as their firms influence policy. Economists' concern for increased efficiency (too many wells demanding unitization, too much flush production losing natural gas, too many wells cutting pressure and ultimate yield) and conservation go unheard as the oil industry more often than not effectively cows the Department of Justice, entrenches itself within the Congress and the Department of the Interior, and establishes its own self-government.[5] Statisticians working for the Bureau of Mines in Interior are effectively reduced to servants for the industry-dominated, quota-setting Interstate Oil Compact Commission.[6] A similar characterization can be applied to the coal industry, in which the economist's or even safety-engineer's rationale only with difficulty penetrates the ongoing policy process.[7]

Urging compulsory unitization or conservation, many scientists have protested through the news media, books, and educational process, but their pleas have not measurably penetrated the policy-making process. They have been "outsiders" unsuccessfully attempting to redefine policy goals, appraise existing policy, and recommend changes. But they have failed largely because they hold no governmental position or have no access to the Executive Branch.

The extractive industries, particularly oil, prefer to keep government out of the research or research-funding business. Thus "federal appropriations for research have always been contested." [8] Robert Engler's incisive and quasi-muckracking analysis of *The Politics of Oil* notes that:

> Industry spokesmen welcome the basic research knowledge gained. . . . But they want the government to get out of this field. Oilmen deplore the conducting of industrial experiments that might easily lead to full-scale operations just as they are wary of any federal research or support that could ultimately threaten their patent-based controls over petroleum.[9]

The oil industry has thus kept a close eye on government-sponsored research into synthetic liquid fuels and fuel production from oil shale and coal. Presumably, they fear government itself developing and profiting from its vast oil shale-bearing Western lands. Some research has already been done within this field in the Bureau of Mines, and extraction processes are now seen as technically feasible. However, even this minimal research was undertaken "despite limited appropriations." [10] The industry would prefer to benefit as much as possible and share as little revenue as possible with the government (cf. off-shore oil and government revenues).

Furthermore, the industry wants to avoid extraction arrangements similar to the atomic energy field where the government buys the ores and has them processed itself in contractually operated (cf. Dow Chemical) laboratories. The likelihood of this arrangement being duplicated for oil is less, however, since atomic energy was (1) new and lacking an established pattern and (2) national security demands were intimately involved.

Extraction policy is shaped in a highly stable arena in which government's role is minimal. There is probably some public support for the actions scientists would desire and encourage, but the support is latent and not demanding. The self-regulative arena is the most hostile and impenetrable to the scientist and his influence.

Extraction industries, including oil, were subject to antitrust actions (Standard Oil) and moderately aggressive administrative regulation in the late 1800s and early 1900s. However, the relationship gradually shifted until government became a promoter and arbiter in a cooperative relationship. This developed when scientists predicted oil shortages, and both industry and government became concerned. Pressures for national security and military reserves (oil for the Navy), conservation, opening of foreign sources (Middle East and South America), and accelerated discoveries west of the Mississippi developed.

The new cooperation gave government its national security reserves. Industry, using government as "enforcer," received price stabilization, protection against oil gluts, and insurance against devastating overproduction. Eventually, tariff and quota arrangements were instituted at the domestic industry's behest to limit foreign competition, once vast fields in the Western United States were opened up. Economists have made some inroads lately on the "uneconomic" character of the 27½ per cent depletion allowance and quotas.

The Justice Department's Antitrust Division has increasingly shown concern with lowering oil prices by reducing quotas. And Congress has had some success, however small, in efforts to reduce the hallowed and traditional depletion percentage. Both developments have been resisted by the industry and its "regulator" friends in the Department of Interior and its Office of Oil and Gas.

But scientists no longer have the rallying cry of "shortage" or "exhaustion." Their sole effective pleas involve economic considerations, and even those meet the determined opposition of "self-regulated" industries. Except for the most determined and patient, this is not a scientist's arena.

THE SOCIAL REDISTRIBUTIVE ARENA

Theodore Lowi notes that conflict in a redistributive arena involves class or politically oriented large interest groups.[11] Demands of "haves" conflict with demands of "have-nots," because the "have-nots" urge a redistribution (as perceived by the "haves") of values. The key values in a social redistributive arena are rectitude, affection, well-being, and respect. Redistribution of wealth may also be explicitly or implicitly involved.

Developed societies, such as the United States, have large middle classes and less numerous minorities of "have-nots." The "haves" dominate these societies both politically and numerically. Thus minority groups now demanding a redistribution of respect, affection, rectitude, and well-being, if not wealth, face a hostile and large majority that does not want what it *has* to be redistributed. These minorities' demands are fragmented geographically, culturally, racially, and ethnically. Their fragmentation encounters an integrated decision system manned and supported by "haves" that will grant certain symbolic benefits (perceived as beneficient distribution) but engage in no large-scale redistribution of wealth or respect.[12]

In the long run, however, issues seen as having redistributive impacts (cf. progressive income tax, welfare payments) may evolve into "distributive" situations. However, the fear of redistribution often lingers. Witness the recent increased popular concern that welfare checks are "redistributive." Thus welfare policy, in its early years, and perhaps even population policy can be treated in part as social redistributive arenas.

Conflict in a redistributive arena is both allocative and structural.[13] The issue, of course, revolves around the allocation of values in a society but also, for many people, raises the question of the proper role of government in society. Working from the premise of a laissez-faire economist or Social Darwinist, many argue that government should not tamper with existing distributions of values, that these distributions are rational and just according to people's abilities and efforts and perhaps according to God's Plan.

Despite the fact that side-payments may render the game nonzero sum,[14] redistribution is basically seen as a zero-sum situation. Thus a high degree of conflict forces conflict resolution and policy making to higher levels in the Executive Branch, often involving the President and Cabinet-level officials. In such polarized "have vs. have-not" situations, a government can respond *only* incrementally, though usually in favor of redistribution to "have-nots."

Government is reluctant to act for redistribution when a vast majority of its support comes from "haves" who perceive proposed government policy to be "redistributive." Scientists' participation, most would agree, usually means recommendations, diagnoses, and pressures implying redistribution. Thus the Executive Branch might itself favor redistribution but be reluctant to shape policy to that end. In this situation, an arena perceived as socially redistributive would have no need for the social scientist who would argue for a redistribution of respect, affection, well-being, or rectitude.

Social scientists' contributions are neither logically or actually always bound to be "redistributive" in impact. Some measures, such as solutions involving increased education as a means to righting wrongs, may indeed be perceived as "distributive" in the American culture. Therefore, social scientists, like the New Economics stressing the creation and distribution of new wealth to all sectors of society, may avoid the stigma of "redistribution."[15] But I am arguing here that they have not generally done so in the past, or at least are perceived as not having done so.

The arena responds more to polemic and class-oriented ideologies than to dispassionate science. Furthermore, the sciences most relevant to the social redistributive arena are themselves suspect as "socialist sciences."

Problems under these conditions are more effectively defined and pursued by the muckraker than the scientist (cf. Michael Harrington's unusual blend in *The Other America,* an attempt to define the poverty problem).[16] Problem definition is crucial in social redistributive policy making. Indeed, should liberals succeed in getting the issues treated as "distributive" (guaranteed annual income, economic rights) rather than as "redistributive," they will find less opposition to the measures they advocate.

Side payments are often means of shifting the issue to a distributive impact. The Tantalus of increased food sales no doubt led retail food store owners and farmers to support food stamp programs. Urban renewal became "distributive" merely by redefining its impact as a means to cleaning up unsightly slums, building new cultural centers, and increasing construction payrolls.

The social sciences and economics are the sciences most relevant to social redistribution. Economics relates to social policy when it argues for a redistribution of wealth *in order to* increase respect, well-being, or rectitude. The influence of social and economic sciences on social policy is rather low within the Executive Branch. Political leadership within the executive branch is somewhat indifferent to their contributions and often divided within itself over policy.

The arena of social redistribution is less developed and stable, unlike the stability inherent in the self-regulative arena. Policy-makers perceive almost no scientific component or criteria entering into the shaping of policy. Thus the need for behavioral scientists as advisers or policy-makers is minimized. The scientists are relegated to access in the low levels of the Executive Branch, and they perform no prescription, invocation, or application functions. Their activities relate more to the educational process, public, and media than to formal policy making. Their influence has been exerted outside more than inside the Executive Branch. These public activities, many would argue, have led to a redefinition of social policy goals, perhaps in some cases shifting people's perceptions from redistributive to distributive impacts. But these activities only indirectly shape the policy positions formed in the Executive Branch.

No sense of urgency augments behavioral scientists' influence, and when a sense of urgency does occur in this arena, as will be shown, policy-makers turn to other skill groups. Policy making is open to public scrutiny and opinion. There is no support and often sharp hostility in the general political climate to proposed "redistributive" policies. Opposition is located throughout the population, and scientists' efforts do not benefit any major private enterprise or vested interest that might support them. The behavioral scientists themselves evidence no desire to work actively in government, nor do they have any vested interest. Most desire social change, but few want to direct or actively pursue it. Some do desire to influence outcomes, but not as policy-makers and often not even as advisers.

John Moeser suggests several factors which might characterize social policy-making.[17] Contrasting the "decisive consensus" behind the space program with the "dissensus" behind efforts to solve the urban problem, he finds that the urban problem:

1. Poses no clear goal or exterior threat
2. Is accompanied by a sense of futility that the problem is insurmountable and "normal" for a society
3. Is a sociological rather than technological problem
4. Has "justice" and "equality" as goals rather than "prestige" and "security"
5. Lacks support and recognition in non-urban areas
6. Presents a complex maze of political, economic and social factors

7. Threatens many citizens, particularly when militants appear to want change in the basic rules of the policy-making game.
8. Requires change in the basic institutions of society which brought on poverty and social problems
9. Lacks a leadership structure, administrative unity (OEO, Department of Health, Education and Welfare, Department of Labor) in Washington, and a basis for concerted elite action.
10. Is not perceived as a problem requiring federal action and programs resulting in conflict with local elites and officials protecting their own realm.

Scientists' participation in a social redistributive arena becomes more clear with an analysis of their role in the formation of *social policy* itself, an area fraught with "dissensus."

Scientists in a Social Redistributive Arena: Social Policy

Social or behavioral scientists either potentially or actively participating in the social policy process face an environment generally inhospitable to policies perceived as redistributing respect, rectitude, affection, or well-being. The environment similarly opposes the efforts of economists who would redistribute wealth in order to redistribute respect, rectitude, affection, or well-being. Such scientists, for these and other reasons, thus achieve only a low level of influence over policy within the Executive Branch.[18]

Social questions and issues are inevitably more controversial and politically sensitive. The injection of a science, which clearly seeks to describe and explain the political or social structure, poses a double threat to the political system —as a challenge to political criteria for solving social problems and as a critical exposure of the existing political system.

Low Levels of Scientist Influence

Many policy-makers feel they have sufficient experience with "people" and their problems that they are themselves competent to make policy without scientific information. And for many the social sciences are not really "scientific"—or at least not as scientific as the so-called hard sciences. The social sciences are seen as less competent sciences in comparison with fields like physics and biology.

Symbolic of policy-makers' self-reliance and disdain for social "science" was a recent dialogue that transpired in a Congressional hearing on a National Science Foundation request for granting funds. Senator Warren Magnuson (D., Wash., Chm., Senate Independent Offices Appropriations Subcommittee) wanted to know the details of various projects to be funded under a $1.5-million allocation for political science, among them a University of Michigan (Survey Research Center) project on the formation of political interests. The dialogue went this way:

> Howard M. Hines (NSF Division Director for Social Science): They are interested in studying to find out how it is, for example, that children get interested in politics and develop their early ideas about politics.
> Magnuson: Are you kidding me?
> Hines replied that he was not kidding.
> Magnuson: I don't quite understand that. I think maybe we can help you out and save $1.5 million.
> Senator Gordon Allott: Or maybe more.
> Senator Allen Ellender: If you go down the list, you will save more than that.[19]

Social knowledge comes from experience and common sense rather than science, many policy-makers would argue.

The social scientist can offer no impersonal, antiseptic "quick technological fix"[20] for social problems, and has fewer quick solutions comparable to contraceptive devices or the automobile. Social change advocated by social scientists or social diagnosis offered by social scientists inevitably must strike a sensitive nerve. People want neither exposure of their selves nor solutions to human problems that require personal involvement or altered styles of living. Witness the storm of personal and political reaction to the Moynihan Report when it became public. Religious groups, the "black" community, Negroes, and the "permanent government" or welfare establishment each reacted differently and strongly.[21] Moynihan's "science" suddenly became not science but political and personal opinion.

Charges that social science could not be scientific because of human independence, too many uncontrollable variables, values, constant change, and people's ideologies, may rise to the surface. And, indeed, there is some real basis for the charge that:

> The fundamental cause of the difficulty is the high degree of indeterminacy in most of social science knowledge. The social scientist acting as advisor to the policy-maker or the administrator cannot often present him with very reliable knowledge about the alternatives he faces.[22]

Social scientists are often insensitive to the needs of policy-makers, who require action-oriented recommendations rather than exhaustive investigations, or justifications rather than a dispassionate balancing of alternatives. Max F. Millikan has covered these different role perceptions elsewhere.[23]

This does not deny that serious academic studies do not have a long-term effect on policy. Some do, but in the short run various factors still reduce the influence of social scientists. Indeed, the long-run and complex effect of many social theories and theorists (cf. Marx, Ricardo, Darwin, Spencer, Keynes) has been significant, but they influenced societies and elites, not just policy-makers in executive branches. Similar comment applies to Gunnar Myrdal's exposé of American racial problems in *An American Dilemma*. Likewise, one can treat the impact of Michael Harrington's impassioned rhetoric in *The Other America*.

Often rhetoric or a raking-of-the-muck can penetrate policy-makers' or elites' minds, while *science* remains trapped in academic circles. Social scientists may not particularly like the possibility, but immediate impact may often demand passion or at minimum the passionate disclosure of reason. The muckrakers and reformers of earlier eras knew this well.

Policy-makers continue to see social problems in political terms, partly because the public generally still sees them in political terms. Thus leaders in the executive branch are often "indifferent" to potential contributions of the social sciences. Indeed, when a sense of urgency develops around a serious social problem, policy-makers are forced by the public to turn to nonscientific skills. Citizens fear social problems that result in riots and civil disorders. But these fears translate into a dimension of law and order or a military response rather than perception of a social problem requiring economic and social scientific solutions. The skills demanded are military rather than scientific. The issue is defined as "law and order," not rectitude or a redistribution of well-being or respect. Scientists cannot provide an immediate means to order, though they might argue that they have a long-run path to that desired end.

The social redistributive policy arena and social policy are undeveloped policy-making situations. This fact is illustrated by two examples. The first is suggested in Gilbert Steiner's *Social Insecurity* when he notes that even social welfare policies have not been based on adequate research.[24] The second is evident in the long time lag between proposals for a Council of Economic Advisers, Economic Report, and the use of "economic indicators" and proposals for a Council of Social Advisers, Social Report, and the use of "social indicators" in policy formation. Even the prototype, *Toward A Social Report*, produced by H.E.W. in January 1969, is but an embryonic effort with an uncertain future.

The primacy of economic security (cf. fear of another depression) forced economists into a significant policy-making role as early as 1946, while the long, slow, and perhaps abortive fight for a Council of Social Advisers did not begin until the late 1960s. Senator Walter Mondale's proposed Full

Opportunity and Social Accounting Act of 1967, Bertram Gross' work with social indicators, and Senator Ribicoff's proposed Office of Legislative Evaluation for the Congress are evidence of the lagging trend.[25] The social sciences are indeed gaining in prominence and influence, but it must be pointed out that the forces that contributed to the Full Employment Act of 1946 were more powerful and compelling than the forces behind the present drive for "social advice."

Indeed, the United States and its Executive Branch have become more aware of social problems, social policy, and social scientists' contributions since 1960, but the arena remains undeveloped for scientists. Part of the answer lies in the arena's "redistributive" character.

THE GOVERNMENTAL REDISTRIBUTIVE ARENA

Samuel Huntington has distinguished *structural* and *strategic issues* involved in defense policy making.[26] Governmental redistribution involves structural issues with basic conflict revolving around the distribution of responsibility, position, and power within a group. But conflict over rules of the game and the distribution of power and responsibility for policy making are the most political of political issues.

Therefore, when the impact of a policy is seen as redistributing power or respect within an organization, opposition to change materializes. Since the economic, systems, and organizational (behavioral) sciences are concerned with organizations and their reorganization, their influence stands to be diminished. It is assumed that scientists' contribution in this arena involves research or advice directed toward reorganization, though sometimes their purpose might well be the sustenance and justification of the status quo. And reorganization implicitly threatens redistribution of the power and respect valued by leaders of the organization who benefit from the existing distribution of these values.

Low Levels of Scientist Influence

The "redistribution" game is zero sum. Some gain what others lose. Leaders benefiting from the status quo can be expected to develop ideologies hostile to change or justifying existing alignments. Groups within an organization might be expected to be reluctant about reorganization, for the change might jeopardize the stabilized relationships they have with one another. Thus, policy making becomes in effect the maintenance of "no-policy" through "nondecisions." The policy is one of no change.

The risks of reorganization are high, since new rules or structures may emerge which will guide future allocations of respect and power within the organization. Furthermore, the new rules may mean a shift in the organization's goals as new individuals and groups come to share in policy making as a result of the reorganization. This is why structural conflict is more fundamental than allocative conflict.[27] This is why much reorganization deals only with objectives such as "economy and efficiency," norms that have minimal political risks and redistributive impacts. This is why similar efforts such as reapportionment or constitutional change often stir so much fundamental conflict.

Scientists concerned with political behavior, organizations, economics of organization, and systems analysis have only low influence within the arena of governmental redistribution. The existing organizational structure and political leadership of the government can be expected to resist their suggestions and advice, confine them to such "safe limits" as economy or efficiency, or not ask their advice in the first place. Reorganization is a political matter, and political criteria must predominate. Science has a minimal role, even though it has formal access through the Presidency (cf. Hoover Commission) and Cabinet levels. But even this access may be token and limited to explicit purposes. This seems to make sense because scientists tend not to hold policy-making positions in respect to prescribing or invoking governmental reorganizations.

The organizational sciences are seen by policy-makers

as less competent than the "hard" sciences. They are seen as relatively easy to comprehend, criticize, or reject. Political or administrative experience serves many policy-makers as a personal reservoir of expertise in reorganizing one's own organization.

There is not much support for scientists' efforts in structuring organizations. Private enterprise encourages and vicariously enjoys any efforts by government to become more efficient and economical (i.e., more like private enterprise) but lends no support beyond Hoover Commission-like activities.

Finally, scientists concerned with the structure and performance of governmental organizations have no driving desire to influence policy making. They have no substantial vested interest to pursue, and many are well aware of their irrelevance for policy making.[28]

Organizations are reluctant to engage in internal redistribution of power and respect. Scientists are welcome only for the achievement of "efficiency" or "economy" and perhaps for minor readjustments, but beyond that they are less welcome. The Executive Branch's own efforts to reorganize and scientists' influence over that reorganization provide a useful example of the governmental redistributive arena.

Scientists in a Governmental Redistributive Arena: Organization Policy

Organization policy encompasses the behavior of the Executive Branch in reorganizing itself and therein potentially redistributing such values as power and respect. If the Executive Branch is reluctant to substantially reorganize, and if scientists attempt to influence that policy, they are bound to meet with frustrations and to have a low level of influence.[29]

Scientific research on organizational behavior and administrative management existed for many years without any measurable impact on government. This condition persisted into the 1945–1968 period, with only a few notable exceptions in specific fields. Explanation for the scientists' stagnant and unused status lies largely in the resistance of large organizations—particularly political or governmental ones—to internal change. Such change threatens to alter existing distributions of power and respect, threatens vested interests within the organization, and increases insecurity in new relationships among all individuals in the organization. Anthony Downs' *Inside Bureaucracy* points out that older organizations have larger proportions of "conservers" hostile to change and that larger organizations are more likely to change their external activities than their internal behavior or organization.[30] Thus age and size mitigate against influence for scientists of organizations, and studies of external functions are more welcome than studies of internal factors.

Regardless, changes in the procedures, rules, or organization of policy-making groups are more fundamental and sensitive than changes in the substance of policy itself. Scientific studies of organizations may be conducted either independently or contracted by government for symbolic purposes, but they are blocked or badly riddled when implementation is tried. Change is accepted when it is minimal, superficial, and does not seriously disturb existing relationships. Perhaps, then, organizational scientists' influence would be greater in the structuring of organizations or bureaus when they are being newly formed.

Governmental reorganization is often a symbolic act, either for the benefit of the electorate or for the psychological benefit of the organization itself. The focus on "efficiency" and "economy" in executive reorganizations illustrates this point. In addition, "economy" and "efficiency" are one step removed from political and positional change and, as such, are less threatening to established relationships and the distribution of respect and power values.

Preoccupation with economic virtue has inhibited organizational scientists' influence. President Taft's Commission on Economy and Efficiency (1910) and the first and second Hoover Commissions on the Organization of the Executive

Branch of the Government are cases in point. Even the new Planning-Programming-Budgeting System (PPBS) being implemented in government departments is significantly justified and bound by the "economy" norm.

The Hoover Commissions did not build on much social science research or the organizational sciences. Commission reports use quotations and arguments from existing authorities and texts for justification and embellishment. Some social scientists served as the consultants and one, Herman Pollack, was a Commission member. But the Commissions were concerned with cutting expenditures, decreasing duplication, consolidating functions, increasing efficiency, and defining functions and activities more clearly. This was the intent of their enabling legislation. Basic direction came from distinguished citizens and civil servants, and much analysis was done within the government agencies themselves.

Such self-analysis has a way of being hopelessly unscientific or at least unobjective. The Commissions made some small use of outside management firms and research organizations. The Brookings Institution wrote the transportation and welfare reports; the National Bureau of Economic Research prepared the statistical agencies report, and the National Records Management Council handled the records management report. However, this process barely tapped the resources of the organizational and social sciences.

Neither the Rockefeller Panel (after 1952) nor the Eberstadt Report (1948) significantly used scientific research on organizational structure and behavior. Scientists experienced in organizational studies were not meaningfully involved or consulted prior to those reports.

Indeed, more success and influence was achieved by reorganization experts prior to the 1945–1968 period. That experience includes Frank Willoughby's Institute for Governmental Research (later the Brookings Institution) and Roosevelt's President's Committee on Administrative Management, though both were seriously limited by existing power and administrative structures.[31] The whole New Deal reorganization experience (PCAM) shows the extent to which organizational experts, relatively free from the norms of "economy" and "efficiency," can threaten the existing distribution of power and respect. Because they posed such a threat, the efforts of Luther Gulick, Charles Merriam, and Louis Brownlow met defeat in the hostile Congress and political climate of the late 30s, even though these men went so far as to push the advocated changes with testimony and lobbying before the Congress. The final reorganization was drawn up in 1939 by legislative leaders. It omitted all the controversial features the experts had proposed and was mild in tone.[32] The attempt to give the President "power commensurate with his responsibility" was partly blunted. It threatened to shake the foundations of the status quo.

Several considerations stem from the organizational sciences themselves rather than larger political factors. Scientists studying organizations have difficulties studying some levels of organization with scientific techniques. The lower administrative levels are best for scientific analysis insofar as behavior there is more routinized and less personal. But policy and significant changes come from the higher levels where, as Selznick has contended, the logic of routine behavior and efficiency loses its force and the logic of leadership (personality variations, individual prerogative) is the rule.[33]

Furthermore, the science is less developed, and perceived by policy-makers as less developed, compared to the so-called hard sciences. Nevertheless, it has become sufficiently mature and "scientific" to pose a threat to "conservers" in organizations. The insights in a concept linking policy and administration, the notion of administrative policy making, a focus on individual behavior in organizations, and a no-holds-barred attitude toward change have simultaneously rendered the scientists more threatening and sophisticated.

But influence remains low, despite evidence late in the 1945-1968 period that scientists are being increasingly consulted by government agencies for evaluation and recommendations. The scientific management studies of the Taylorites, the decision-making theories of Herbert Simon, and the human relations approaches of Kurt Lewin, John Dewey, and Elton Mayo were little used or consulted by government.[34]

Chris Argyris' recent work in the State Department on communications behavior is an example of an invitation to study and consult. But Argyris' criticism of rigidities within the State Department is not likely to alter the existing organization. Nevertheless, a scientist has penetrated the initial stages of the organization policy process. And organizational scientists, like scientists generally, are bound to be used more in an increasingly complex and technological society. But relatively, they may remain less influential than other scientists.

THE EXTRA-NATIONAL ARENA

The impacts of governmental policies in the "extra-national" arena are considered to be *"redistributive,"* though many notable groups in the United States would prefer to see the impact as distributive and many more others are totally unaware of any impact at all. Redistribution takes either wealth, respect, or security from the United States and gives the wealth, respect, or security to nondomestic groups and foreign nations.

Foreign aid, seen in its early years as "distribution," came to be considered as a "redistribution" of wealth to enemies and neutrals.[35] Disarmament and arms control efforts, for many, threatened to redistribute security—lowering the United States' capabilities and increasing the net advantage of enemies. Diplomacy (foreign political policy) has been conducted so that the United States has sacrificed its self-respect to increase the respect given to other nations.

The basic issues in the extra-national arena are more allocative than structural and are perceived as zero-sum games. Many argue that disarmament and arms control are nonzero-sum games, that nations' security increases mutually with

96 Low Levels of Scientist Influence

such measures, but these voices are often lost midst pleas for military superiority. But allocation is not always the policy objective. Membership in, and redistribution of "sovereignty" to, the United Nations and foreign alliances —allowing them to be independent entities, which can commit or act against the United States or her interests—constitute a threatened structural change, altering the pattern of international policy making and international authority.

But whether the redistribution is structural or allocative, the perceived impact is similar—domestic harm or loss of wealth and respect, increased national insecurity, almost no domestic material benefit (except increased exports through foreign aid), and minimal domestic symbolic benefit in the sense of "giving" to those who need.

The arena is fraught with conflict. The immediate benefactors of governmental activities, whether distributive or redistributive, are nondomestic groups that have no votes for Congressmen or Presidents. At "best," government action in the extra-national arena is altruistic philanthropy. At "worst," it is bribery. Domestic benefits are indirect (increased exports, military allies), and most Americans see more direct ways to increase security (military superiority), wealth (trade), and respect ("big stick diplomacy").

The arena taps an "insecurity" complex in the American character. Policies are seen to sap the United States' national strength. One problem with America's experience in Vietnam has been the "extra-national" character of her policy in that area. Involvement has been defended with altruistic motivations or often simply on the grounds of resisting Communists. But the War has neither been conducted as a full scale effort nor successfully justified as a defense of vital U.S. security.

Insecurity for Americans results from nonmilitary solutions to problems and less-than-total commitments of her military power. Thus the insecurity and frustration of Vietnam. The impact of the United States' involvement in Vietnam has come to be seen as "extra-national" and redistributive, unlike World War II when involvement directly met domestic needs and increased Americans' communal security.

The formation of extra-national policies is decentralized but involves the gamut from the President down to agency levels. The policy-making structure is stable and highly tradition-bound. The arena belongs to diplomats who value their own expertise and experience and who maintain their own professional tradition and autonomy.

Scientists who attempt to influence policies in the extra-national arena must immediately confront these bounds of tradition and diplomats' own self-reliance in policy making. The policy-making structure is thus relatively impermeable, and scientists do not hold formal positions where they might officially recommend, prescribe, or administer policy. They are confined to (1) research and (2) involvement, access, or influence through the media, public, and educational process. The role of mobilizing public opinion through the media is most apparent in scientists' disarmament and arms control campaigns.

Political leadership in the Executive Branch could be characterized as "indifferent" to scientists' contribution, except insofar as their technology influences arms control negotiations (black boxes, detection of underground devices) or their leadership of public opinion narrows policy-makers' freedom of action.

The sciences involved run from the social sciences (diplomacy, foreign political policy) and economics (foreign aid) to biology, physics, and the engineering sciences (disarmament and arms control). Economists and the "hard" or engineering sciences have been more successful than social sciences. Their influence (foreign aid, disarmament and arms control) has been moderately low, while social scientists' influence has been low (diplomacy, foreign political policy). Generally, the arena is a "wasteland" for scientists wishing to influence policy or see their product significantly used.

Policy in the arena depends only minimally on scientific components. Policy-makers can more easily comprehend most of the science they confront, and with the exception of disarmament technologies they take a dim view of the scientists' competence and the "scientific" nature of their science. A significant amount of political conflict among scientists over objectives, strategies, and values may be apparent *and* detrimental to their influence. Conflicts over scientific issues are also apparent.

No strong sense of urgency compels policy-makers to use scientists' contributions. Except insofar as they might want influence or funds for research in the area, scientists themselves have almost no significant existing vested interest to pursue. In fact they encounter and are frustrated by vested interests such as the State Department's traditional monopoly over diplomatic affairs and the vaunted "military-industrial-political complex."

Some scientists and their science form a *peace industry* in the extra-national arena, even when they are working on detection devices relevant to disarmament or arms control. Their interests are not the immediate interests of the "warfare" or "garrison" states. And no general support backs or encourages these scientists' efforts in influencing policy. Still, this may constitute a vested interest which might mature.

These generalizations about the extra-national arena mask several interesting differences and circumstances that surround individual policies within the arena. Therefore, attention should be given to specific policy types such as foreign political (diplomacy), foreign aid, and disarmament and arms control policies.

Scientists in an Extra-National Arena: Foreign Political Policy

Policy is considered "foreign political" insofar as it encompasses the normal range of diplomatic activity not directly related to major crises, military questions, or foreign aid. The central values being distributed or redistributed are respect, rectitude, and affection. The State Department is the major policy-making institution, dominating a highly stable policy-making process during the entire 1945-1968 period.[36]

Zbigniew Brzezinski has noted a lack of scientific and academic research devoted to foreign political policy making. He observes that:

> Foreign policy is perhaps the last important area of organized activity in the United States that still operates largely on the basis of combining the intuitive judgment of a few individuals with the traditional thrust of bureaucratic inertia generated by a large professional organization.[37]

A Stanford Research Institute report to the Senate Foreign Relations Committee urged in 1959 that the State Department devote more attention to the relevance of social and psychological sciences for policy making, but no major changes have resulted from that recommendation.[38] Indeed, the White House Office and the Department of Defense harbor more experts on foreign affairs than the Department of State's Bureau of Intelligence and Research and Policy Planning Council (Staff).

The State Department itself lets no contracts for "outside" research. The Department's Office of International Scientific and Technological Affairs (Science Office) went without leadership recently for a few years, leaving a void of advice on the impact of science on international relations.

Research funds and scientific advice are concentrated in two fields—international scientific activities and arms control —and neither of these areas is significantly endowed. Arms control policy is considered separately within the extranational policy arena. Even there, however, Rene Dubos *vividly points out an interesting fact*

> One dramatic illustration of this negligence is the research budget of the State Department. Science, lavishly endowed by public funds, produced nuclear weapons—the means by which man can now destroy himself. The problem of pre-

venting this catastrophe is primarily the State Department's responsibility. Indeed, there is very little support for any kind of scholarly work on the explosive international issues now facing the world.[39]

Furthermore, it is not clear that research for the Arms Control and Disarmament Agency, beyond that on technologies of detection, has any measurable impact on policy in an arena so sensitive to political criteria and demands.

The State Department does give significant attention to international scientific activities, but these efforts do not provide scientific inputs for policy making. They involve policy for science and scientific exchange rather than "science for policy making." Eugene Skolnikoff finds five science-related activities in the foreign policy process, but this research deals directly only with one—the neglected service that science and scientists can perform for policy making. Skolnikoff characterizes such a function as "issues and opportunities that grow out of the application of techniques and analytical tools of the sciences to foreign policy problems." [40]

Barriers to the use of science and scientists in foreign political policy making are thrown up by scientists, policy-makers, and the subject matter itself. Scientists and scholars have not been enthusiastic about working with or for the State Department. Many suspect its institutionalization of bureaucracy, shun its "isolationist" tendencies, and feel uncomfortable within such a tradition-laden institution. These inhibitions pertain particularly to the social scientists, those scientists most relevant to the policy being studied. And it is the social scientist who represents the scientific community in foreign political policy making.

There is a common fear abroad that social and physical scientists are being "used" for political ends. The social sciences, furthermore, are somewhat undeveloped as sciences for the study of foreign political matters. Often scholars talk a language incomprehensible to policy-makers, inquire on a plane too general and theoretical for policy-makers, and prefer to study "unimportant" aboriginal tribes rather than the more "relevant" USSR.

The subject matter itself poses difficulties. Alternatives cannot easily be tested and verified in advance, experimental opportunities are lacking, world situations change all too frequently, international actors are often erratic and have wills and secrecy of their own, and information is generally less reliable.

The conditions and personnel shaping the policy-making process present additional difficulties. Observers have suggested that a:

> . . . strong traditional "humanist" bias against the social sciences and perhaps the sciences in general, probably contributed to the State Department's failure to develop any significant program of research support.[41]

Foreign policy-makers apparently prefer to make policy by relying on experience and intuition. They will encourage and use descriptive and background analysis, but go no further in either wanting or using scientific explanations or predictions. They want only the facts and will use intuition and experience to forge explanations and expectations. Skolnikoff argues this point, noting that the:

> . . . validity of the results of systematic study in social science subjects is not necessarily seen by many operators in the field or in Washington to be superior to judgments based on experience or common sense. Moreover, the problems of relating the often imprecise results or highly qualified projections of most social science research to operational needs are frequently great.[42]

Foreign policy-making requires generalists, not specialists writing exhaustive studies, the policy-makers argue. There is a need for action-oriented, timely, precise, and advice-oriented science. But science is explanation-oriented, often ambiguous, and untimely. Brzezinski argues that:

102 Low Levels of Scientist Influence

> The intellectual malaise of much of contemporary American social science reinforces the intellectual timidity of American thought on international politics. Fearful of daring generalization, preoccupied with method and relying on quantification, American scholarship on international affairs provides useful tools only to those policy-planners whose ambition is to refine but not to create, to elaborate but not to perceive, to conserve rather than change.[43]

Professor Brzezinski therefore argues for a Council of International Affairs, comparable to the Council of Economic Advisers and reporting to the President. Still, social science is seen by policy-makers as neither scientific nor useful, despite its attempts to be or appear scientific or useful.

The conditions of policy-making force scientific advice into the background. The immersion of foreign service officers in daily routine, pressures of time on decisions, work overloads, de-emphasis of anticipatory planning and actions, pressures for consensus and compromise with other nations and one's colleagues in State, preference for internal over external advice, minimal research budgets, and a hesitancy to challenge basic assumptions further aggravate the policy-maker's feeling that science is irrelevant to policy making. Furthermore, foreign policy often develops in a "charged" and emotional domestic environment, a fact itself imposing constraints on reason, planning, creativity, open-mindedness, and the use of science.[44]

Foreign-area research has become a sensitive activity in international politics. The Project Camelot episode, though associated with the Department of Defense, illustrates the problem. The realities of the situation probably will mean less, and perhaps less-valid, research, though both problems might be partly alleviated were such research dissociated from the military and located in the Department of State.[45] This would solve another problem at the same time by decreasing State's dependence upon intelligence and technical agencies outside the Department. Currently, officials must get information and scientific advice from the President's Science

Advisory Committee, Office of Science and Technology, Atomic Energy Commission, and Department of Defense. They have no way of checking on that advice.

The State Department consults intermittently with experts on foreign areas and international affairs, but this sporadic and often non-research-related consultation does not elevate scientists' influence on policy beyond a low level. Science's role remains largely undeveloped. This applies to the use of scientific studies in policy making *and* to the use of scientific techniques or technologies in decision making (cf. Planning-Programming-Budgeting System). Explanations for the less-developed science and policy relationship, as has been shown, relate to policy-makers, scientists, conditions of policy making, and the often inscrutable subject matter of foreign affairs.

Scientists in an Extra-National Arena: Disarmament and Arms Control Policy

Arms control and disarmament measures are still generally perceived to have "redistributive" impacts. Other nations' are seen to gain security at the United States' expense when arms control or disarmament occurs. Some argue that arms control and disarmament produce increased communal security for citizens of the United States, but that view does not yet predominate in the formation of policy. This situation has persisted since 1945. The period under analysis (1945–1968) harbors, as will be evident, a stable policy process in a relatively unchanging political situation.

Arms control and disarmament refer to arrangements between nations and involves a process that shapes or limits their use of weapons in international life. Scientists' participation in this policy type is substantially different from their role in foreign aid, weapons, deterrence and defense, and normal foreign policy formation.

A specific discussion is in order since this is a special type of scientist involvement in a policy field with elements of foreign and military policies. The scientists' influence on disarmament and arms control policy can be described as *moderately* low, despite a high level of policy and influence-oriented activity on scientists' part.[46]

Scientists have had influence despite a plethora of constraints, including a public fearful of arms control or disarmament, uncompromising United States' foreign policies vis-à-vis Communist bloc nations, and the difficulties of negotiating with other nations on arms control or disarmament measures. Their influence has stemmed from their role as *initiators* of most shifts in United States' nuclear weapons policy since 1945. The lone exception was Secretary of State John Foster Dulles' massive retaliation doctrine.[47]

But scientists' initiative has rarely carried over into other functions in the policy process, beyond lobbying and mobilizing public opinion behind measures scientists initiated. Therefore, the inner circles of the policy process have been without scientist participation. Scientists have been innovators of ideas, leaders of public opinion, advisers, technological experts, and developers of technologies, but not policy-makers or administrators.

INITIAL USE OF THE ATOMIC BOMB. The Franck Report, written by Eugene Rabinowitch and expressing the thoughts of Leo Szilard and James Franck, was the initial major effort by scientists to influence arms control. The Report, dealing with the immediate issues surrounding the use or nonuse of the newly developed atomic bomb, expressed an appreciation of the new character of international relations under atomic umbrellas. Such understanding was not then apparent in higher political circles of the government.

However, the scientists urging nonuse or a demonstration of the bomb failed, largely because their argument was based on nonscientific and political grounds and because a group of scientists formally advising the key Interim Committee advocated military use of the weapon. The decision "could have been altered only if scientists had been fully united."[48] They were not united nor would they be united in subsequent attempts to influence policy.

POST-WAR DOMESTIC AND INTERNATIONAL ATOMIC CONTROLS. Transformed from strategists to citizens during the postwar demobilization, many scientists, particularly those involved in atomic energy, realized they had a new obligation. They developed a moral commitment in new directions, deeming the *use* of their discoveries as important as initial discovery itself. Critics saw this subsequent engagement in public debate and lobbying as "meddling" in the political arena without competence or portfolio, but the scientists had a different perspective.

Albert Einstein's proposal to President Roosevelt had been made out of social obligation to insure that the Allies would not be defeated by a German *wehrmacht* equipped with nuclear weapons. This same sense of social responsibility, now turned inward, explained the scientists' pleas for a simple demonstration of the atomic bomb, in the hope that the awesome prospect of a real attack would bring the Rising Sun to its knees. It further explains the postwar drive for civilian control (McMahon Act, Atomic Energy Act creating the AEC) and international controls (Baruch Plan) of atomic materials and weapons. This new willingness to engage in public debate manifested itself in the columns of the *Bulletin of the Atomic Scientists,* a morally impelled and policy-oriented publication. Scientists had moved into open debate and assumed leadership of public opinion.

Scientists were nominally successful on domestic control insofar as they seized on a sensitive issue (civilian control) and defeated the May-Johnson Bill and substituted the McMahon Act. The real issue might not have been civilian control, but the scientists were able to define it that way and win. They had no organized opposition, and scientists who disagreed or had reservations about the substitute McMahon Bill kept silent. Therefore, scientists were not seen as divided. Furthermore, they had access within the Executive Branch and Congress at many points and were "assisted" by a clumsy legislative campaign waged by the military (bans on open discussions, total secrecy on plans, perfunctory hearings on the May-Johnson Bill, misleading assertions of scientist support, disregard for other Executive Agencies and Departments).

Though scientists won a formal victory, the resultant "civilian" AEC was for years preoccupied with military requirements. This may be evidence of the strength of basic political demands stressing security through superiority and arms rather than through parity and arms control.

Initiative for the Baruch Plan itself did not come from scientists, but the drive for international controls did. Actual drafting of the Plan was done within the State Department under Secretary of State Dean Acheson's direction. Coaching on the substance of the plan came from scientists Vannevar Bush and J. Robert Oppenheimer. The Federation of Atomic Scientists' contacts with the State Department "were certainly not decisive and can scarcely have accomplished more than to underline the support that existed among scientists for a system of control based on inspection." [49]

However, this role in preparing the political climate for international controls must not be underestimated. Scientists led the way, even though they were operating in a political capacity rather than furthering scientific knowledge. The Baruch Plan did fail eventually, but perhaps because it had made unrealistic demands, inserted by the State Department, and perhaps because the Soviet Union felt it was in her own self-interest to reject the Plan. The idea failed in a political environment that scientists could not control.

DEVELOPMENT OF THE H-BOMB AND DETERRENCE. Scientists, badly divided, failed to shift U.S. arms policy from 1947 through 1950. As a result, scientists such as Edward Teller and Ernest Lawrence successfully argued for development of the hydrogen bomb, contending that it met U.S. objectives of arms superiority and a competitive attitude in the arms race.

The arguments of probomb and antibomb scientists were founded on their professional, political, and emotional characteristics as well as their respective views of American priorities and Soviet intentions. The decision to build a bomb was made over the advice of the Atomic Energy Commission's General Advisory Committee (scientists) and J. Robert Oppenheimer. Teller's arguments were successful despite significant scientific conflicts about the bomb's feasibility. Generally, therefore, one could expect that scientists can be scientifically or politically divided and the "hawk" scientists' position prevail. But scientists cannot be divided if the "dove" or antidevelopment scientists are to win. The H-Bomb made Dulles' massive retaliation doctrine possible, and "hawk" scientists (cf. the Teller group) predominated in policy making through 1957.

FALLOUT, PARITY, MUTUAL DETERRENCE, AND TEST BANS. The fallout issue was raised largely by scientists. "Hawk" scientists, nominally grouped under Edward Teller as chief spokesman, had dominated executive policy circles through Sputnik (1957). After 1957, however, the Eisenhower Administration gradually began to get competing scientific advice with the creation of the President's Science Advisory Committee and the introduction of new scientific voices—Rabi, Kistiakowsky, Killian, Weisner, and Zacharias. These new voices urged mutual deterrence, efforts toward arms control, and a test ban. Even Dulles was:

> . . . so impressed with the newcomers' depth of scientific knowledge and their authority, which obviously outclasses almost all of their opponents, that he admitted he had been given poor scientific advice before the new group arrived.[50]

Still, United States' initiative on a test ban depended upon the extent to which suspension of tests would imperil the

108 Low Levels of Scientist Influence

nation's position in the arms race. Superiority rather than parity was the predominant policy position. This constraint of "military security" has hung over arms control initiatives since 1957 and has inhibited influence by scientists favoring arms control measures.

The 1957 President's Science Advisory Committee group (Bethe Panel), appointed by President Eisenhower at James Killian's suggestion, had no seismologist among its members. Though a majority of the Panel opposed a cessation of testing, they soon became convinced that continued testing by the USSR would allow the Soviets to catch up with American nuclear capabilities. However, Atomic Energy Commission Chairman Strauss and Assistant Secretary of Defense Quarles argued that the military advantage of future American testing could not be forgone. The Panel agreed. Finally, the Soviet Union unilaterally suspended its own testing and the Bethe Panel recommended similar U.S. action in response, but only *after* completion of a current American series of tests.

Pro-test-ban scientists now had access to the White House. The Atomic Energy Commission (AEC) responded by conducting its own tests and concluding that existing detection devices would not successfully detect explosion, even in the air. The pro-ban scientists advising Eisenhower countered the AEC argument, contending that they were not depending on *existing* technology. This group felt that a new technology, one expressly designed for detection of nuclear explosions and not earthquakes, could and would be developed. However, antiban scientists erected further barriers. RAND physicist Albert Latter suggested to Edward Teller that underground explosions in cavernous "big holes" could muffle an explosion or make it resemble an earthquake. This "decoupling theory" forced omission of underground tests from the 1963 Moscow Treaty, even though technology soon made their detection likely.[51]

Scientists' participation in test ban negotiations ranged from status as diplomats (Conference of Experts, Geneva,

1958) to technical advisers (Moscow Treaty, Geneva, 1963). Prior to the test ban efforts, scientists' participation had been explicitly political. But the fallout arguments and detection technologies brought scientific judgments to the forefront, and the Executive Branch needed technical advice. Prior scientific debates had dealt with the use or demonstration of the atomic bomb (probability of failure in detonation) and development of the hydrogen bomb (feasibility, costs, and cutbacks in atomic weapons program to free materials for H-Bomb project). With the test ban effort, scientific criteria again applied to detection devices, means of evasion of detection, and future weapons innovations. The scientists, having contributed toward a political climate in which a test ban could be sought, stepped into a *limited role* as technical experts serving a political objective (test ban) selected and pursued by nonscientist politicians and diplomats. They had laid the groundwork and would then provide technology and technological forecasting.

Scientists' roles as diplomats and negotiators at the 1958 Geneva Conference of Experts is generally deemed to have been unsuccessful, though observers differ over who is to blame for the failure.[52] As the test ban negotiations proceeded until 1963, scientists were limited to advisory functions, particularly as the final treaty was being shaped in 1963. They provided background data for the diplomats, were used to mobilize public support, and were relied upon to interpret technologies to laymen. Similar roles for scientists may be associated with the Strategic Arms Limitation Talks (SALT) recently pursued by the U.S. and USSR in Helsinki and Vienna.

Beyond this narrow role, many scientists, themselves unfamiliar with existing or future detection technologies, spoke out on the advisability of a ban. Their views were based on different "strategic" positions. Teller and John S. Foster (then Director, Lawrence Radiation Laboratory, Berkeley) led the "hawks," while Linus Pauling spoke as a "dove." Both groups capitalized on their status as scientists to shape opinions through the media. But the Kennedy Administration's determination to achieve a test ban—a position based on political rather than scientific considerations—was the major factor. Scientists were only window dressing to impart a technological and scientific aura which might justify the action.

Scientists favoring arms control or disarmament measures for scientific or strategic reasons face a hostile environment not encountered by more "hawkish" scientists. Political reality more readily encourages and tolerates an arms race and efforts for military superiority.[53] Indeed, the later discussion of the weapons policy process will confirm this judgment.

Furthermore, prospects for arms control must confront an influential, though subtle, amalgam of interests in a "weapons culture," [54] "warfare state," or "garrison state." Such coalescences of interest unite the military, congressmen, weapons manufacturers, labor unions associated with defense contracts, local economies, and even scientists themselves working on weapons systems. Many scientists, normally predisposed toward arms control and efforts for "peace," doubtless find themselves unconsciously drawn into the emerging "military-industrial complex." In this manner, arms control silently yet surely loses spokesmen. Even scientific organizations are drawn into the "culture" and no longer have a critical capacity or independence for action.

Jerome Weisner, contending that "disarmament systems are actually more complex than weapons systems," [55] argues that overcoming the weapons culture will require men:

> . . . to create a vested interest in arms control; to develop a cadre of people whose full-time occupation is research and development of means of arms control and on the analysis of the political and military problems of arms control.[56]

Senator Frank Church, noting that the Arms Control and Disarmament Agency (ACDA) had a total budget of only $400,000 in Fiscal 1960 (.001% of the Defense Budget), observed, "I think it may be fair to say on the basis of those figures that there is hardly any room at the inn for peace."[57] ACDA's research budget allows almost no scientific inquiry into arms control and disarmament, and even the research that is done is frequently disregarded or buried out of deference to the sensitive political climate in which ACDA operates. The Agency is a political instrument, and political considerations have penetrated its internal use of research.

So research and governmental awareness of arms control and disarmament remain in a primitive state, both politically and scientifically. ACDA has no capability for technical research into detection devices and must depend on the AEC or Department of Defense, two horses of a different color, who might not share ACDA's interests.

Even research on the economics of disarmament, let alone studies of the politics of disarmament, remains undeveloped. Some spillover of knowledge beneficial to arms control has developed from Defense Department work (cf. Thomas Schelling's work on bargaining, RAND studies for the Air Force) but not nearly enough to place the sciences studying ways to peace on a par with sciences studying ways to wage war.

Arms control and disarmament policy has been shaped in an atmosphere lacking a sense of urgency. No pressing demands encourage consultation and deference to scientists' judgment. The scientist's participation remains undeveloped. So often he must wage his campaign in the media as a "political" figure with a scientific reputation. He lacks both formal position and deference in policy making within the Executive Branch. Much of his participation involves the expression of political, strategic, or value preferences rather than scientific judgments.

Often the scientists turn to political or strategic argument because they lack scientific evidence, or fear becoming embroiled with existing scientific conflicts. Scientists lack a reservoir of arms control-related science and may quite rightly be seen as less scientifically competent on many questions.

They have at best been "used" as advisers when their "advice" tends to support the political objectives of the user. But this should not obscure the larger contribution of the "atomic scientists"—creating a climate in which there is at least a willingness to consider arms control measures, even if meaningful steps in that direction are impeded by a more powerful preoccupation with military superiority and military means to security.

Scientists in an Extra-National Arena: Foreign Aid Policy

Foreign aid activities have been perceived as both distributive and redistributive in the postwar (1945–1968) period.[58] However, the recent cuts in foreign aid budgets have resulted from an increasing tendency in Congress and public opinion to see aid as a "redistribution" of wealth and respect to other nations. The early years of foreign aid programs had been less conflict-ridden and more distributive in impact.[59] But recently, scientists' efforts to influence aid policy have been increasingly frustrated. The Executive Branch, occupied with retrenching on foreign aid programs, has been relatively less concerned with innovation in types of aid, more economically sound criteria, and shifts to multilateral means. Aid administrators doubtless agree with the scientists' objectives (more aid) but have less use for the scientists and their science, insofar as they contribute to policy making. The arena harbors too much hostility for a developed scientist-policy relationship.

The Agency for International Development does indeed hire scientists but primarily as consultants working on specific aid programs for specific countries or regions. These might include scientists from the fields of agriculture, economics, and the engineering sciences, but such individuals do not in general influence or attempt to influence overall aid policy.

Economists, rather involved and influential when their field intersects with other fields of policy making, fare poorly when attempting to influence foreign aid policy. This poor relationship has existed in a stable fashion since 1945. The United States has funded research on foreign aid problems since the Agency for International Development (AID) was created in 1961. The earlier record shows less support. The first AID administrator, Fowler Hamilton, placed research far down on the list of priorities, and it has remained there ever since. And priorities become important in an agency already pinched on funds.

Research also suffers from an "indifference" not unlike the attitude of foreign-policy makers in the State Department, where intuition and experience rather than science shape decisions. An abortive attempt to rescue research on agricultural and public health problems was made by David Bell as AID administrator, but old-line bureau chiefs blocked his efforts. No significant funding has developed since then, and research further lagged when political support for foreign aid declined through 1968.

Political criteria, short-run "crisis" needs, and less-scientific considerations shape much aid policy. And what planning or comprehensive policy-making is attempted is frequently disintegrated by these more immediate and politically powerful pressures.

Foreign aid has no domestic clientele or constituency. Its support lies abroad, coming only indirectly from domestic circles when the support of industry is enlisted with "side payments" to join the pro-aid coalition. The stimulation of industry's exports by aid dollars returning to the United States (a high percentage are spent here) is such a "side payment" or benefit. Research, being a part of legislative allocations for aid, suffers when aid is cut. Furthermore, the domestic and foreign political insecurities that accompany an aid program may tend to result in administrative timidity in funding potentially controversial or innovative research. Administrators would not want scientific findings or questionable projects to rock an already sinking boat.

Research also suffers from farmers' hostility. They would

oppose research on agricultural development in other nations simply because it ultimately is intended to increase foreign agricultural productivity and perhaps threaten some American food exports.[60] But private enterprises other than farming may benefit in the short and long run, through increased markets and trade, both of which should compensate for some loss of markets to newly developing foreign operations.

Economists studying development in foreign nations face a difficult scientific problem in multitudinous and uncontrollable variables. Their "science" has not yet developed to a satisfactory level and is therefore less useful in making current aid policy.

Finally, scientists' activities in influencing policy have been confined to the agency level and to the public media. They have had no influence over aggregate funds committed to aid; the various percentages allocated to allies, neutrals, and members of hostile blocs; aid cut-offs to nations seizing American property without compensation; and priorities among internal functions within AID itself. But the economic sciences have shaped individual aid decisions and refined criteria for aid grants. Therefore, their influence is characterized as only *moderately* low.

THE DISTRIBUTIVE ARENA

The impact of government's activities is often perceived and accepted as "distributive." Government is seen distributing benefits to groups who do not see themselves in competition with one another. Logrolling and pork barrel situations typify this arena and indicate its basic allocative rather than structural character.[61]

One group's claims are not treated as affecting another group's claims. The game is perceived by Congress, the Executive Branch, and beneficiaries to be nonzero sum. Indeed, these private groups, distributing executive agencies, Congressional committees, and interests supporting private groups often coalesce into an amalgam living off the distribution of benefits to the private group.

Groups with uncommon interests thus make independent demands on the government and together with Congress and an agency agree on a procedural rule of mutual noninterference with other groups' demands. Therefore, as Salisbury and Heinz would view the situation, fragmented demands are confronting and receiving benefits from a fragmented decision system, since government itself grants agencies, bureaus, and committees autonomy to distribute without mutual interference.[62] The arena, like logrolling processes, is highly stable and persists relatively unchanged from year to year, with only new groups and demands entering the distribution process.

The arena often borders on being a self-regulative arena, particularly when government allows private groups to distribute benefits themselves among their own subunits or write their own rules (cf. NRA codes, some aspects of agriculture policy). Distributive policy may also resemble entrepreneurial policy when government consciously fosters an increase in some value (cf. science policy, research funding, scientific knowledge) but becomes closely intertwined with the fostering process. Thus, space and atomic energy would be entrepreneurial arenas, but science policy might easily be treated as "distributive" policy.

Government takes no aggressive role as an initiating, independent actor in distributive policy making. It merely reacts to demands from private or other governmental groups. At most it may enter into a cooperative arrangement with a private group (cf. government's role as financial backer and design-approver for the SST). But, in general, government responds to private demands and gives legal sanction to meeting them.

Labor policy, educational policy, urban renewal, housing policy, water resources policy, and pollution control policies are often *distributive*. Personnel policy for parties and governments is often a matter of distribution. Trade and tariff policies from the 1920s to late 1950s were distributive in their impact.

Often, however, what appears to be "regulation" is merely disguised "distribution." When the Civil Aeronautics Board grants a route to United and denies that route to American Airlines, it has made a choice. But it may try to compensate later with a new route for United, therein balancing government benefits and insuring that no major carrier is in financial jeopardy. This is allocation and distribution with an intent to balance benefits despite the impossibility of two groups winning in any given situation. Furthermore, what formally passes for regulation may often be a clientele-captured regulatory agency distributing wealth, power, and respect to the groups it is supposed to regulate. Government in such cases is not an actor making real choices or depriving private interests.

When government is merely distributing benefits to private groups, it does not need scientists. Thus their influence in a distributive arena will be moderately low. Having no need to defend its position and no burden to develop science itself, the Executive Branch may use scientists primarily to assure minimum quality in funded programs allocated to private groups. It can merely distribute funds to private groups which in turn develop their own scientific advice and science.

Scientists' relationship with the Executive Branch assumes an *advisory* character, though particularly in agriculture policy scientists do engage in government-funded research in government laboratories. They do not hold policymaking positions and thus neither prescribe, authorize, apply, or administer policies. But the political leadership of the Executive Branch does need and encourage their contributions to policy making. Furthermore, that leadership is neither divided nor indecisive. The direction of government policy is established and stable.

The relevant sciences usually will have rather clear "applied" interests, such as those of the life sciences (biology, chemistry) in agriculture and of the system and engineering sciences in transportation. Specialization and a reputation for competence no doubt contribute to scientists' influence, even as advisers and researchers. Scientific conflict is minimal, and scientists could be described as having a significant vested interest in the applied fields they are developing. Their success and future are tied to the future of transportation and agriculture and to technological progress in those areas. These interests may range from patents to employment, contracts, research support, and profits on investments. This is likely whenever the sciences have significant "applications," particularly in physical technologies.

Private enterprise, whether agricultural or industrial, benefits from scientists' involvement as consultants or researchers associated with the Executive Branch. Thus, private interests want the scientists involved in the Executive Branch because they are developing scientific knowledge that will be distributed by the government to the private interests. There may be no general support for scientists' participation or government distribution, but each benefiting group has a high level of interest in government's activities, and this is sufficient.

Scientists' participation in the Executive Branch will be clearer with a brief but selective treatment of two "distributive" policies—transportation and agriculture.

Scientists in a Distributive Arena: Transportation Policy

The power behind transportation policy-making resides in the Departments of Transportation, Housing and Urban Development, and Health, Education, and Welfare. Policy outputs consist of grants-in-aid, highway construction funds, research and development contracts, regulation of routes and fares of commercial carriers, and guidelines for new transportation designs. The types of scientists, whose influence can be described as moderately low, include economists, systems analysts, planners, and engineers. The policy they attempt to influence deals with highway, rail, pedestrian, and air transportation for intraurban and interurban needs. The description of transportation policy-making has in this case been limited to the 1960–1968 period, because prior to 1960 policy dealt almost entirely with automobiles and highway construction, except for the on-going "regulatory" agencies dealing with rail and air fares and routes.[63]

Scientists' participation includes (1) economists gathering data for regulatory agencies; (2) engineers, economists, and systems analysts advising administrators on the distribution of grants-in-aid; (3) transportation engineers engaging in in-house and contract research on new modes of transportation; and (4) planners and systems analysts developing long-range, innovative transportation plans for the nation.

Three conditions negatively affect scientists' influence within the Executive Branch:

1. *The Lure of the "Open Road."* Policy-makers' preoccupation with highways for the automobile, though substantially less since the early 1960s, limits the activities and innovative efforts of scientists concerned with transportation. Though scientists' efforts to refine and improve highway transportation would be welcomed by a "highway-automobile" bloc, they are limited in other areas by the highway interests. It is understandable that automobile manufacturers, oil companies, highway contractors, state highway departments, suburbanites desiring to drive to work rather than ride, and the Bureau of Public Roads have a vested interest in preventing the development and encouragement of new modes of mass transit and systems of propulsion (cf. electric cars). Nongasoline modes of transportation surely evoke opposition from petroleum interests and auto manufacturers.

Still, some manufacturers have developed and are developing interests in new modes of transportation (cf. aircraft companies, manufacturers of buses and other mass transit systems). Transportation engineers and economists would welcome these new options. But most Americans evidence a distaste for mass transit and believe their automobiles have social significance. Exceptions might include some rail commuters in major cities and subway riders in New York.

2. *Distrust of Federal Involvement and Planning.* The "highway bloc," and citizens generally, have distrusted federal involvement in these traditional state and local responsibilities. Given this aversion, federal expenditures for the Interstate Highway System were better "justified" by national defense needs.

The antiplanning norm so prevalent in American politics has permeated transportation policy to some extent, rendering transportation activities "stop-gap" in nature and inhibiting an anticipatory role for government in meeting future transit problems. Thus scientists' skills are that much less-needed in the Executive Branch. Transportation engineers, economists, planners, and systems analysts share common irrelevance in policy making.

3. *Regulatory Agencies with "Distributive" Impacts.* Economists and systems analysts have some formal relevance for the Interstate Commerce Commission, Civil Aeronautics Board, Federal Aviation Administration, and other agencies with regulatory functions over transportation. However, often these agencies do not really regulate and instead distribute benefits to their clientele. When this is the case, scientists should not be needed, because the agencies are not active combatants in a game denying private groups their wishes. Of course, decisions on routes and fares cannot please all carriers at any given point in time, but over a period of years a setback on one fare increase or route authorization can be compensated with remedial action. Or a regulatory agency's disapproval of a request may later, when the "symbolic" value of "regulation" is not needed, be reversed. Regulatory agencies are filled with politics and political considerations. Scientists would find them less-than-ideal places to be influential.

Distributive arenas are *political* arenas. But politics does not always mean hostility to scientists. Often it can open new

120 Low Levels of Scientist Influence

avenues for influence and research. If the Executive Branch does not need scientists within any on-going distributive policy process, it could need them should the extent of distribution expand into new fields. Thus the expansion of "distributive" activities into urban mass transit, supersonic transports, and new automotive technologies can mean (1) new research opportunities supported by government, (2) new fields in which political leaders need information and advice to satisfactorily distribute benefits, and (3) an emerging role advising government on the relative amounts of benefits it will distribute to the varied types of transportation. The last role would involve the skills of economists and systems analysts, but its full development would require government to actively *regulate* the behavior and benefits accruing to one mode of transportation versus another mode. But that hour has not yet been reached. Regulation is still seen as a process of choosing within a sector (air transportation, rails, oil, etc.) and *not* as a process of choosing between competing sectors (drug companies vs. consumers, railroads vs. commuters, highways vs. airways). Both processes can be regulative, but choice between competing sectors is more difficult and fraught with political trouble for an Executive Branch. The difficulties of "regulating" *within* a sector are themselves rather forbidding.

Urban political power has been rising, and national problems have been given increasing attention. Transportation problems are primarily "urban" and "national" in scope, and thus a new constituency has risen to combat the traditional "highway" bloc. These urban problems have been more the bailiwick of the Executive Branch than of the Congress, as urban interests have rallied around the Presidency. Michael Danielson explains that

> The main sources of support for positive federal involvement continued to come from central-city political leaders and Eastern commuter railroads. In 1960, planners, metropolitan transportation experts, and urban specialists from the academic community were added to the ranks.[64]

But transportation policy develops within a distributive arena. When government is distributing benefits, scientists' primary contribution and influence comes from their ability to develop new technologies and "hardware" that can be distributed. Indeed, pressures from the clientele demand this assistance from the Executive Branch, either in the form of research funds or the results of research.

But there are limitations. The transportation industry, when associated with mass transit, wants "hardware" that has immediate payoff and utility. Such demands do not in the least encourage research into total new concepts in transportation but tend to benefit engineers more than scientists. Speeding up existing railroad facilities with faster trains and rails is an answer but may be no more than a stop-gap measure. Such technological changes become further misdirected when one realizes that a more difficult and pressing problem occurs within, rather than between, metropolitan areas.

Scientists' influence in a distributive arena depends largely on their contributions as discoverers and inventors of new technologies and knowledge. Scientists who do not participate as providers of new technologies for distribution to private groups are not needed in distribution politics and have minimal influence.

CASE STUDY: THE SUPERSONIC TRANSPORT. Scientists have made the SST possible, but they have also been amazingly successful in slowing down its development and eventual production. Physicists and sound engineers have raised the issue of noise pollution from sonic booms that may be produced by the plane. Thus it is generally agreed that the plane will be used only on over-water routes, or at subsonic speeds on domestic routes.

Economists have argued that the plane's effect on the United States' balance of payments might be nullified. Proponents of the craft argued that its sales abroad or the threatened sale of Anglo-French Concordes in the U.S. would justify its production to save the U.S. from increased balance-of-payments problems. Some economists felt, however, that increased foreign travel by Americans would be stimulated by such faster transportation and effectively cancel out any balance-of-payments advantages.

Initial economic studies of the plane favored its development, but airlines themselves are increasingly wary of the load factor and cost problem. The carriers already have excess capacity and fear that the new "jumbo" jets (Boeing 747) will lower fares so far that the SST could not compete with its higher fares and faster service. Many economists have also argued that, in effect, the government's subsidy to Boeing for developing the plane taxes everyone for the benefit of the 15 per cent who fly.[65]

But the SST's major supporters—its contractors and the Federal Aviation Administration, which "commissioned" its development—remain determined, despite the recent slowdown on development. Still, the "case study" reveals the possibility that scientists can influence distributive policies from positions "outside" the Executive Branch. They can have an impact when their scientific analysis indicates that a program is (1) not feasible or impossible, (2) likely to generate intolerable side effects, or (3) so uneconomic as not to be worth the investment. Indeed, Professors Stockfisch and Edwards have argued in *The Public Interest* that had not the government substantially bankrolled Boeing's development of the SST, it might never have been built:

> At a minimum, if the government adopted the financial and management plan proposed here, it would "smoke out" the true belief in the private manufacturing sector as to whether the program will be viable. It is our bet that the revealed behavior of the private sector would suggest that either the SST program should be dropped, or that it should proceed at a much slower pace.[66]

Still, it is a tribute to the powerful political character of distributive policy that scientists, even when they identify serious problems with a program or government-distributed project, may not ultimately change the outcome. Planes, dams, and weapons systems are more often than not built in spite of scientific objections. Only the rationale—profits for industry, votes for the local congressman, and military invulnerability—differs.

Scientists in a Distributive Arena: Agriculture Policy

Agriculture has long been the beneficiary of a clientele-oriented public policy. In effect government has *distributed* wealth, enlightenment, respect, power, and skills to the farming community. Most sectors of the society have tolerated this distribution of benefits and accepted the action as part and parcel of distributive politics. Nothing has fundamentally changed in this arena since it first developed, and certainly not since 1945. If anything, the arena has become more "self-regulative" in some respects.

Scientists and agriculture policy-making have a long-standing relationship, despite an abundance of conflict over policy goals and means of implementing those goals.[67] However, a scientist's success in influencing the policy process depends on his field of endeavor. Chemists, botanists, zoologists, and other life scientists have enjoyed a long, mutually beneficial association with policy-makers and policy, while economists have met some frustrations and social scientists generally have had only a minimal amount of influence, despite their work in agricultural extention programs.

But scientists who develop new technologies and "hardware" are bound to be more relevant in a distributive arena, where government can merely support their research as a "subsidy" to the clientele's interests. The life scientists have provided benefits for the clientele (increased production per acre, new strains of plant and animal, new production hardware), while economists and social scientists have often threatened "regulation" or a "redistribution" of values—contributions quickly seen as not beneficial to the clientele. The efforts of the latter have meant more than applications-oriented science and government "assistance." And distribution arenas are *political* arenas, shaped by political forces and clientele demands rather than scientists.

Chemists and life scientists have been influential because the benefits they produce (productivity, freedom from plant disease, new planting methods) have been a clear benefit to the farming industry. But their activities have not gone beyond their immediate fields of animal husbandry, plant genetics, pest control, soil improvement and maintenance, hardware. Only economists have seriously sought to influence the larger scope of agriculture policy, and they have met its "political" character head-on.

Chemists and life scientists flourished in the Department of Agriculture, university colleges of agriculture, and Soil Conservation Service (SCS) "because no single farmer has ever been able, as have some industrialists, to subsidize scientific research." [68] This relationship has persisted despite serious conflicts. Biologists have been concerned with pesticides and fertilizers and their effect on streams after run-off, while the Extension Service scientists apparently consider the problem as a simple need to increase crop production. Colleges of Agriculture purportedly battle with the Soil Conservation Service, charging SCS with a narrow erosion preoccupation while the Colleges feel production is a much wider and appropriate focus. These same "cow colleges" enjoy significant *de facto* independence, in part due to alliances with major farmer groups like the Farm Bureau. SCS has further been criticized by economists for treating symptoms rather than diseases. The economists argue that erosion is not the problem, citing instead economic instability of farming, unfavorable land-to-man ratios, rural poverty, and insufficient types of proper credit.[69] Finally, the Production and Marketing Association (PMA), an organization run by farmers, though formally within the Department of Agriculture, serves as a pro-parity pressure group on the Secretary of Agriculture. PMA deems farming experience rather than agricultural research as the sole background for leadership.[70] But despite these factors, chemists, life scientists, and agricultural economists do live off the agricultural complex and have a significant vested interest in its growth and problems.

Economists' participation has produced varying influence over policy. Their most significant role, however, occurred during policy's early years and the New Deal period. Robert Eyestone has identified three cycles in agriculture policymaking. Using Lowi's scheme, he labels them "redistributive," "regulative," and "distributive" but adds a recent fourth —"self-regulative." [71]

The early years of policy were seen to have a "redistributive" impact; i.e., the period running from the New Deal through 1947, during which price supports and controls became a reality. Economists during the New Deal successfully encouraged an end to laissez-faire treatment of farming and, with Roosevelt's support, convinced farmers of the merits in parity policies. Sociologists, psychologists, and anthropologists were also involved in the effort, but to a lesser degree. The economists provided analysis of the greater decline in agricultural prices vis-à-vis industrial prices, an analysis giving the farmers an argument for their self-interests. These economists drew up the initial legislation, lobbied for its passage, and therein succeeded in establishing the concept of parity, retiring submarginal land, gaining acceptance for government intervention even in a "distributive" manner, and raising farm incomes.

The economist's influence finally met a roadblock when they moved beyond "distributive" measures and urged land-use planning, lower consumer prices in cities, economic planning, improved status for the rural poor, and shifts in production (cf. less reliance on cotton in Southern states). The political structure surrounding agriculture, particularly in Congress, and rising "conservative" farm groups, notably the American Farm Bureau Federation, came to distrust government efforts toward *planning* and the economists' close relationships with the Secretary of Agriculture. The feeling emerged that scientists should stick to science and avoid making policy, particularly policy with "redistributive" and "regulative" impacts.

By 1940 the previously influential economists saw their influence begin to wane.[72] Furthermore, a new Secretary of Agriculture took office, war preoccupied the economy, and people became suspicious of recent and rapid changes in society. But more fundamentally, farmers wanted higher prices (distribution) and economists wanted a large-scale reorganization and planning for agriculture (redistribution, regulation). Obviously, the clientele and its political muscle prevailed. Economists' efforts were restricted. Even in the 1960s Wallace Sayre described the environment for science in the Bureau of Agricultural Economics as "too severe for survival." [73]

Economists have recently argued that price supports, which they encouraged in the 1930s, are uneconomic. But here too the political character of agriculture policy enables victory for those with a vested interest in current policy.[74]

The New Deal redistributed wealth to the farming industry (parity, rising farm incomes), but the arena gradually became distributive and self-regulative. Attempts at formal regulation occurred from 1948–1959, but in a larger sense government was still distributing benefits to agriculture.

Encouragement of scientists is confined to research activities of immediate, tangible and perceptible value; and scientists' attempts to intrude on larger policy-making activities meet rebuffs from the farming clientele. The clientele sees the scientist as a servant and resists any efforts he might make either to become or to lead government to become an active combatant in policy making (regulation, redistribution).

So political is the agriculture policy process that even scientific research and development can meet resistance. The famous oleo debates, involving nutritional, economic, and health arguments for oleo's production, is an excellent case in point.[75] Similar legal and political obstacles surround recent controversy about synthetic "milk" products. Agriculture belongs to the politician and farmer.[76]

Notes

1. The concepts of "structural" and "allocative" policy types come from Robert Salisbury and John Heinz, "A Theory of Policy Analysis and Some Preliminary Applications," *Policy Analysis in Political Science,* ed. by Ira Sharkansky (Chicago: Markham Publishing Co., 1970), pp. 39–60. A similar distinction between "structural" and "strategic" issues, relative to defense policy, is developed in Samuel P. Huntington, *The Common Defense* (New York: Columbia University Press, 1961), pp. 4–6.

2. Robert H. Salisbury, "The Analysis of Public Policy: A Search for Theories and Roles," *Political Science and Public Policy,* ed. by Austin Ranney (Chicago: Markham Publishing Co., 1968), p. 159.

3. *Ibid.,* p. 168.

4. Murray Edelman, *The Symbolic Uses of Politics* (Urbana: University of Illinois Press, 1967), pp. 23–29, 36.

5. Robert Engler, *The Politics of Oil: A Study of Private Power and Democratic Directions* (Chicago: University of Chicago Press, 1961), pp. 372–394. For an economic argument, consult Wallace F. Lovejoy and Paul T. Homan, *Economic Aspects of Oil Conservation Regulation* (Baltimore: Johns Hopkins Press for Resources for the Future, Inc., 1967). A more recent and dispassionate account of oil policy making is Gerald D. Nash, *United States Oil Policy* (Pittsburgh: University of Pittsburgh Press, 1968).

6. Merle Fainsod, Lincoln Gordon, and Joseph C. Palamountain, Jr., *Government and the American Economy* (New York: W. W. Norton and Company, Inc., 1959), pp. 664–679.

7. *Ibid.,* pp. 621–638.

8. Engler, *op. cit.,* p. 98.

9. *Ibid.*

10. *Ibid., op. cit.,* p. 96.

11. Theodore J. Lowi, "American Business, Public Policy, Case Studies, and Political Theory," *World Politics,* XVI : 4 (July, 1964), p. 713.

12. Salisbury, *op. cit.,* p. 168.

13. Salisbury and Heinz, *loc. cit.,* and Huntington, *loc. cit.*

14. Salisbury, *op. cit.,* pp. 158–159.

15. I have developed this elsewhere, linking it to the difficulties surrounding recommendations requiring changes in human behavior and life style. Dean Schooler, Jr., "Political Arenas and the Contributions of Physical and Behavioral Sciences and Technologies to Policymaking" (Paper delivered at the Sixty-Fifth Annual Meeting of the American Political Science Association, New York, September 2–6, 1969). A revised and shortened version appears as "Political Arenas, Life Styles, and the Impact of Technologies on Policymaking," *Policy Sciences,* I : 3 (tentative, Fall, 1970).

16. Michael Harrington, *The Other America: Poverty in the United States* (Baltimore: Penguin Books, 1963).

17. John Moeser, "The Space Program and the Urban Problem: Case Studies of the Components of National Consensus," Staff Discussion Paper, Program of Policy Studies in Science and Technology, George Washington University, Spring 1969.

18. Major sources on social scientists and social policy include Lee Rainwater and William L. Yancey (eds.), *The Moynihan Report and the Politics of Controversy* (Cambridge: M.I.T. Press, 1967); Raymond A. Bauer (ed.), *Social Indicators* (Cambridge: M.I.T. Press, 1966); Daniel Bell, "The Idea of a Social Report," *Public Interest,* 15 (Spring, 1969), pp. 72–84; Mancur Olson, Jr., "The Plan and Purpose of a Social Report," *Public Interest,* 15 (Spring, 1969), pp. 85–97; Department of Health, Education, and Welfare, *Toward A Social Report* (Washington: Government Printing Office, January, 1969); Robert J. Samuelson, "Council of Social Advisers: New Approach to Welfare Priorities?" *Science,* CLVII (July 7, 1967), pp. 49–50; Andrew Kopkind, "The Future Planners," *New Republic,* CLVI : 8 (February 25, 1967), pp. 19–23; Fred R. Harris, "Political Science and the Proposal for a National Social Science Foundation," *American Political Science Review,* LXI : 4 (December, 1967), pp. 1088–1095; Edward A. Shils, "Social Science and Social Policy," *Philosophy of Science,* XVI : 3 (July, 1949), pp. 219–242; Max F. Millikan, "Inquiry and Policy: The Relation of Knowledge to Action," *The Human Meaning of the Social Sciences,* ed. by Daniel Lerner (New York: World Publishing Co., 1967), pp. 158–180; Gunnar Myrdal, "The Relation Between Social Theory and Social Policy," *British Journal of Sociology,* 4 (1953), pp. 210–242; and Thomas F. Pettigrew and Kurt W. Back, "Sociology in the Desegregation Process: Its Use and Disuse," *The Uses of Sociology,* ed. by Paul Lazarsfeld *et al.* (New York: Basic Books, 1967).

19. "Social Sciences' Support Problems," *Science*, CLX (May 3, 1968), p. 518.

20. Alvin M. Weinberg, "Can Technology Replace Social Engineering?" *Bulletin of the Atomic Scientists*, XXII : 12 (December, 1966), pp. 4–8.

21. Rainwater and Yancey, *op. cit.*, p. 332. This edited volume is an excellent case study of the role and influence of social science and scientists on policy making.

22. Bernard Barber, *Science and the Social Order* (London: George Allen and Unwin, Ltd., 1953), p. 255.

23. Millikan, *loc. cit.*

24. Comment on Steiner's observation and the use of social research generally can be found in U.S. Congress, House, Research and Technical Programs Subcommittee of the Committee on Government Operations, *The Use of Social Research in Federal Domestic Programs*, 90th Cong., 1st Sess., 1967, I–IV.

25. The best sources on "social indicators" and the proposed Council and Report are Bauer, Samuelson, Kopkind. See Note 18.

26. Huntington, *loc. cit.*

27. Salisbury and Heinz, *loc. cit.*

28. An interesting account of one adviser's awareness of his irrelevance and real role occurs in Harvey Wheeler, "The Short and Happy Life of a Research Consultant," *Western Political Quarterly*, 13 (September, 1960), pp. 852–857.

29. The major sources of information on organization policy making are Paul Y. Hammond, *Organizing for Defense* (Princeton: Princeton University Press, 1961); Richard Polenberg, *Reorganizing Roosevelt's Government: The Controversy over Executive Reorganization* (Cambridge: Harvard University Press, 1966); G. Homer Durham, "An Appraisal of the Hoover Commission Approach to Administrative Reorganization in the National Government," *Western Political Quarterly*, II (December, 1949), pp. 615–623; Charles Aikin, "Task Force: Methodology," *Public Administration Review*, IX : 4 (Autumn, 1949), pp. 241–251; Herman Finer, "The Hoover Commission Reports," *Political Science Quarterly*, LXIV : 3 (September, 1949), pp. 405–419. On a conceptual basis, Yehezkel Dror, *Public Policymaking Reexamined* (San Francisco: Chandler, 1968), is useful in discussing what he calls "policy making knowledge" as distinguished from "policy issue knowledge." These tools for making decisions and policy I would call "technologies" of policy making.

30. Anthony Downs, *Inside Bureaucracy* (Boston: Little, Brown and Company, 1967), pp. 191–210. Downs' brief article elsewhere

illustrates secondary, nonscientific functions of advice and is useful here. Downs, "Some Thoughts on Giving People Economic Advice," *American Behavioral Scientist*, IX : 1 (September, 1965), pp. 30–32. Policy-makers often use and need advice for purposes other than making the "right" decision or assessing alternatives.

31. See Hammond, *op. cit.*, pp. 87–92 for Willoughby's experience and Polenberg, *op. cit.*, for the PCAM experience.

32. Polenberg, *op. cit.*

33. Philip Selznick, *Leadership in Administration: A Sociological Interpretation* (Evanston, Illinois: Row Peterson and Company, 1957), pp. 3–4.

34. Amitai Etzioni, *Modern Organizations* (Englewood Cliffs, New Jersey: Prentice-Hall, 1964), pp. 29ff; James G. March and Herbert A. Simon, *Organizations* (New York: John Wiley and Sons, 1958), pp. 12–33.

35. Theodore J. Lowi, "Making Democracy Safe for the World: National Politics and Foreign Policy," *Domestic Sources of Foreign Policy*, ed. by James N. Rosenau (New York: Free Press, 1967).

Concern with foreign policy as an "issue-area" subject to analysis in terms of redistributive, distributive, and regulative impacts also occurs in James N. Rosenau, "Foreign Policy as an Issue-Area," *Domestic Sources*, ed. by Rosenau (New York: Free Press, 1967), pp. 11–50, and Stephen J. Cimbala, "Foreign Policy as an Issue Area: A Roll Call Analysis," *American Political Science Review*, LXIII : 1 (March, 1969), pp. 148–156. Cimbala finds that foreign policy issues with "domestic" implications tend to be seen as "redistributive."

36. Sources on foreign political policy making include Raymond Platig, "Research and Analysis," *Annals of the American Academy of Political and Social Science*, Vol. 380 (November, 1968), pp. 50–59; W. Phillips Davison, "The Use of Sociology in Foreign Policy," *The Uses of Sociology*, ed. by Paul F. Lazarsfeld et al. (New York: Basic Books, Inc., 1967); Zbigniew Brzezinski, "Purpose and Planning in Foreign Policy," *The Public Interest*, 14 (Winter, 1969), pp. 52–73; Eugene B. Skolnikoff, "Scientific Advice in the State Department, *Science*, CLX (November 25, 1966), pp. 980–985; Irving Kristol, "American Intellectuals and Foreign Policy," *Foreign Affairs*, XLV : 4 (July, 1967), pp. 594–609; C. Wright Mills, *The Power Elite* (New York: Oxford University Press, 1959); Eugene B. Skolnikoff, *Science, Technology and American Foreign Policy* (Cambridge: M.I.T. Press, 1967); U.S. Congress, Senate, Subcommittee on Government Research of the Committee on Government Operations, *Federal Support of International Social Science and Behavioral Research*, 89th Cong., 2nd Sess., June 27-28 and July 19–20, 1966; Herman Pollack, "Science, Foreign Affairs, and the State Department," *Department of State Bulletin*, LVI (June 19, 1967), pp. 910–917. Lincoln P. Bloomfield, "The Political Scientist

and Foreign Policy," *Political Science and Public Policy,* ed. by Austin Ranney (Chicago: Markham Publishing Co., 1968), pp. 179–196; and James N. Rosenau, "Moral Fervor, Systematic Analysis, and Scientific Consciousness in Foreign Policy Research," *Political Science and Public Policy,* ed. by Ranney (Chicago: Markham Publishing Co., 1968), pp. 197–236.

37. Brzezinski, *op. cit.,* p. 73.

38. Donald W. Cox, *America's New Policy Makers: The Scientists' Rise to Power* (Philadelphia: Chilton Co., 1964), p. 88. It should further be noted that the Policy Planning Council under Walter W. Rostow in the Kennedy Administration did have some minimal successes, such as counter-insurgency strategies and "regionalism" as a focus for political and economic development programs.

39. Rene Dubos, "Science and Man's Nature," *Science and Culture,* ed. by Gerald Holton (Boston: Houghton Mifflin Co., 1965), pp. 251–290.

40. Skolnikoff, *loc cit.* The five categories are 1) issues associated with dominant technical objectives (space, atomic energy, peaceful uses of the atom), 2) issues of a political nature but heavily dependent on scientific considerations (ocean utilization, national security), 3) issues of a political nature but not sensitive to technical factors in the short run (disposal of agricultural surpluses), 4) issues associated with the application of the scientific method to policy processes (simulation, war games, survey analysis), and 5) issues associated with the implications of future scientific and technological developments (weather modification, territorial control, genetic control, capabilities for decentralized decision making).

The research here considers survey analysis, war games, and simulations to be social scientists' contributions to policy and policy making as well as more traditional academic studies of foreign areas and politics. It is argued that traditional studies as well as contemporary techniques, such as simulation and war games, are equally less relevant for policy-making.

41. Bruce L. R. Smith, *The Rand Corporation* (Cambridge: Harvard University Press, 1966), p. 318. The point has also been argued by Gabriel Almond (see U.S. Congress, Senate, *op. cit.,* p. 114).

42. Skolnikoff, *op cit.,* p. 204.

43. Brzezinski, *op. cit.,* p. 58.

44. The predispositions of scholars to deviate from the immediate interests of policy-makers in the State Department may diminish scholars' influence. Anthropologists would prefer to study obscure groups while policy-makers are interested in major powers. Political scientists prefer comparative analysis, but policy-makers need data on individual countries. Historians would rather study past opinions and traditions

that have minimal impact on current opinions and situations. These arguments are made by William R. Polk in Irving Louis Horowitz (ed.), *The Rise and Fall of Project Camelot* (Cambridge: M.I.T. Press, 1967), p. 243.

45. The fragile character of research support is nowhere more evident than in the Project Camelot affair, though even here Army and military sponsorship could not survive the storm of criticism stirred in Chile and the United States. The project, initiated by the Army's Special Operations Research Office, affiliated with American University, was not sought by scholars. However, they were left with no support when Chilean Communists raised the issue of intervention or at least possible intervention in Chilean affairs. The research itself was vulnerable because it presumed revolution to be "pathological" and sought to find means of avoiding internal wars in developing lands. Camelot was sacrificed to save the SORO operation, but it did lead to questioning about the proper role of social science research and the propriety of locating it in Defense rather than State. Furthermore, the revelations led to inquiries into Defense's Project Revolt on French Canada, Project Michelson on American, Soviet, and Chinese goals; and Project Simpatico in rural Columbia. Consult U.S. Congress, Senate, *op cit.* and Horowitz, *op. cit.*, Horowitz's introduction (pp. 3–44) is particularly valuable.

46. Sources dealing with scientists and disarmament and arms control policy include Fred J. Cook, *The Warfare State* (New York: Macmillan Co., 1962); Emile Benoit and Kenneth Boulding (eds.), *Disarmament and the Economy* (New York: Harper and Row, 1963); Jerome B. Weisner, *Where Science and Politics Meet* (New York: McGraw-Hill, 1965); Ralph E. Lapp, *The New Priesthood* (New York: Harper and Row, 1965); Harvey Brooks, "The Scientific Adviser," *Scientists and National Policy-Making*, ed. by Robert Gilpin and Christopher Wright (New York: Columbia University Press, 1964); James E. Dougherty and J. F. Lehman, Jr., *Arms Control for the Late Sixties* (Princeton: D. Van Nostrand Co., 1967); Donald A. Strickland, "Scientists as Negotiators: The 1958 Geneva Conference of Experts," *Midwest Journal of Political Science*, VIII : 4 (November, 1964), pp. 372–384; Donald G. Brennan (ed.), *Arms Control, Disarmament, and National Security* (New York: George Braziller, 1961); John W. Spanier and Joseph L. Nogee, *The Politics of Disarmament: A Study in Soviet American Gamesmanship* (New York: Frederick Praeger, 1962); Cecil H. Uyehara, "Scientific Advice and the Nuclear Test Ban Treaty," *Knowledge and Power*, ed. by Sanford A. Lakoff (New York: Free Press, 1966); James R. Newman and Byron S. Miller, *The Control of Atomic Energy* (New York: McGraw-Hill, Inc., 1948); Paul J. Piccard, "Scientists and Public Policy: Los Alamos, August–November, 1945," *Western Political Quarterly*, XVIII : 2 (June, 1965), pp. 251-262; U.S. Congress, Senate, Subcommittee on National Policy Machinery of the Committee on Government Operations, *Organizing for National Security: Science, Technology, and the Policy Process*, 86th Cong., 2nd Sess.,

April 25–27, 1960; Robert Gilpin, *American Scientists and Nuclear Weapons Policy* (Princeton: Princeton University Press, 1962); Harold K. Jacobson and Eric Stein, *Diplomats, Scientists and Politicians: The United States and the Nuclear Test Ban Negotiations* (Ann Arbor: University of Michigan Press, 1966); Eugene B. Skolnikoff, *Science, Technology, and American Foreign Policy* (Cambridge: M.I.T. Press, 1967); Alice Kimball Smith, *A Peril and a Hope: The Scientists' Movement in America: 1945–1947* (Chicago: University of Chicago Press, 1965); William R. Nelson, "Case Study of a Pressure Group: The Atomic Scientists" (Unpublished Ph.D. dissertation, Department of Political Science, University of Colorado, 1965); Philip Noel-Baker, "Science and Disarmament," *Impact of Science on Society*, XV : 4 (1965), pp. 211–246; American Assembly, *Arms Control: Issues for the Public* (Englewood Cliffs, New Jersey: Prentice-Hall, Inc., 1961).

47. Noel-Baker, *op. cit.*, p. 225.

48. Alice Kimball Smith, *op. cit.*, p. 71.

49. *Ibid.*, p. 226.

50. Saville R. Davis, "Recent Policy Making in the United States Government," in Brennan, *op. cit.*, p. 386.

51. A full account of the "decoupling theory" debate is in Cook, *op. cit.*, pp. 202–259.

52. The three different interpretations are argued in Gilpin, *op. cit.;* Jacobson and Stein, *op. cit.;* and Strickland, *op. cit.*

53. Spanier and Nogee's concept of the "joker" further reveals the fundamental political nature of the arms control environment, since insofar as nations insert unacceptable provisions in proposals to force their opponents to reject the proposal and suffer a propaganda defeat, then that much less serious are they about real progress toward control. Spanier and Nogee, *op. cit.*

54. Ralph E. Lapp, *The Weapons Culture* (New York: W. W. Norton and Co., 1968).

55. Weisner, *op. cit.*, p. 183.

56. *Ibid.*, p. 176.

57. U.S. Congress, Senate, *Organizing for National Security*, p. 316.

58. Some sources on scientists and foreign aid policy making are Eugene B. Skolnikoff, "Birth and Death of an Idea: Research in AID," *Bulletin of the Atomic Scientists*, XXIII : 7 (September, 1967), pp. 38–40; Eugene B. Skolnikoff, *Science, Technology, and American Foreign Policy* (Cambridge: M.I.T. Press, 1967); Theodore Geiger and Roger D. Hansen, "The Role of Information in Decision-Making

on Foreign Aid," *The Study of Policy Formation*, ed. by Raymond Bauer and Kenneth Gergen (New York: The Free Press, 1968), pp. 329–380; Michael Kent O'Leary, *The Politics of Foreign Aid* (New York: Atherton Press, 1967); and Joan M. Nelson *Aid, Influence, and Foreign Policy* (New York: Macmillan Company, 1968).

59. Lowi, "Making Democracy Safe for the World."

60. Skolnikoff, *Science, Technology, and American Foreign Policy*, p. 62.

61. Salisbury and Heinz, *loc. cit.*

62. *Ibid.*, p. 3.

63. Sources on scientists and transportation policy include George Lardner, Jr., "Supersonic Scandal," *New Republic*, CLVIII : 2780 (March 16, 1968), pp. 13–17; Robert Burkhardt, *The Federal Aviation Administration* (New York: Frederick A. Praeger, 1967); George M. Smerk, *Urban Transportation: The Federal Role* (Bloomington: Indiana University Press, 1965); and Michael Danielson, *Federal-Metropolitan Politics and the Commuter Crisis* (New York: Columbia University Press, 1965). Louis Kohlmeier's treatment of transportation "regulation" indicates how "distributive" such policy really is. His sophisticated piece of "muckraking," filled with examples, also discusses fields like natural gas, trade, energy, banking and investment, labor, and television. Kohlmeier, *The Regulators: Watchdog Agencies and the Public Interest* (New York: Harper and Row, 1969), pp. 136–170.

64. Danielson, *op. cit.*, p. 138.

65. Lardner, *op. cit.*

66. J. A. Stockfisch and D. J. Edwards, "The Blending of Public and Private Enterprise: The SST as a Case in Point," *Public Interest*, 14 (Winter, 1969), pp. 108–117. The authors argue that government, in absorbing financial risks of development, distorts the market mechanism that should determine the plane's profitability. They argue that earlier investors (i.e., government) in a risky venture should be guaranteed a higher rate of return on their investment. Furthermore, it is contended that the *basic political character* of the SST decision demands that a subsidy be given in order to compensate for an economically questionable commitment. The quotation appears on p. 117.

67. Sources on scientists and agriculture policy include Charles M. Hardin, *The Politics of Agriculture* (Glencoe, Illinois: Free Press, 1952); Hardin, "The Bureau of Agricultural Economics Under Fire: A Study in Valuational Conflicts," *Journal of Farm Economics*, XXVIII : 3 (August, 1946), pp. 635–660; Byron T. Shaw, "Research Planning and Control in the United States Department of Agriculture: The Experience of an Old and Well-established Research Agency," *Annals of the American Academy of Political and Social Science*,

CCCXXVII (January, 1960), pp. 95–102; Murray R. Benedict, *Farm Policies of the United States 1790–1950* (New York: Octagon Books, 1966); Richard S. Kirkendall, *Social Scientists and Farm Politics in the Age of Roosevelt* (Columbia: University of Missouri Press, 1966); A. Hunter Dupree, *Science in the Federal Government* (Cambridge: Belknap Press of the Harvard University Press, 1957); Reynold M. Wik, "Science and American Agriculture," *Science and Society in the United States,* ed. by David D. Van Tassel and Michael G. Hall (Homewood, Illinois: Dorsey Press, 1966); Walter W. Wilcox, "Social Scientists and Agricultural Policy," *Journal of Farm Economics,* XXXIV : 2 (May, 1952), pp. 173–183; and Ernest G. Moore, *The Agricultural Research Service* (New York: Praeger, 1967).

68. Bernard Barber, *Science and the Social Order* (London: George Allen and Unwin, Ltd., 1953), p. 172.

69. Hardin, *The Politics of Agriculture,* pp. 102–103.

70. *Ibid.,* p. 125.

71. Robert Eyestone, "The Life Cycle of American Public Policies: Agriculture and Labor Policy Since 1929" (Unpublished paper given at the 1968 Annual Meeting of the Midwest Political Science Association, Chicago, May 2–4, 1968).

72. Kirkendall, *op. cit.*

73. Wallace S. Sayre, "Scientists and American Science Policy," *Scientists and National Policy-Making,* ed. by Robert Gilpin and Christopher Wright (New York: Columbia University Press, 1964), p. 110.

74. Carl Kaysen, "Model-Makers and Decision-Makers: Economists and the Policy Process," *The Public Interest,* 12 (Summer, 1968), p. 83.

75. Don K. Price, *Government and Science* (New York: New York University Press, 1954), p. 103. The Iowa oleo case involved dairy interests demanding dismissal of a research assistant at an Iowa university for publishing findings on the production costs and nutritional value ratios between margarine and butter during a period of wartime shortage.

76. The "distributive" character of agriculture policy making is illustrated by the continuing subsidy for tobacco farmers, when another arm of the government is attempting to "regulate" cigarettes. Political demands force the government in opposite directions simultaneously, but we have learned to accept this duality and ambiguity.

The Policy-making Process and Moderate Levels of Scientist Influence

CHAPTER 5

Scientists' influence can be described as *moderate* in the *regulative* policy-making arena. Policies in this arena may be described as domestic and "regulative" in their impact. Regulation involves the *prohibition or prescription of behavior by the government* or its Executive Branch. Government is perceived as having made *real, tangible choices* between implicitly or explicitly competing private interests.

Government becomes an intervening actor that asserts its own preferences in social choices and enforces those preferences. It may deny renewal of a broadcasting license, enforce automobile safety standards it has drafted, defend the interests of consumers against producers, or decide which carrier will get what routes. It may regulate competing interests within sectors (airlines, broadcast media, industry) or divergent interests between sectors (con-

NOTES TO CHAPTER 5 START ON PAGE 163

sumers vs. producers), but it must *regulate*. This means that later *compensatory behavior,* effectively rendering government activity "distributive" in impact, cannot occur. An agency or commission cannot deny a route to one carrier, award it to a second, and six months later give preference on a new route to the carrier initially losing a decision. Government's active intervention must be meaningful and involve actual *choice,* though more often than not these criteria are not fulfilled.

Scientists' influence rises to a moderate level when government's activities are perceived to have a "regulative" impact, primarily because government then needs scientists and their science. If an Executive Branch is planning to act in a manner that will be perceived by the interested private parties as "regulative" and as government intervention, then the Executive Branch, aware of this reaction, will need scientists to develop technologies, provide scientific justification and rationale for the government's position, and predict the reactions of the "regulated" parties that would be deprived or indulged.

Government might want cheap and functioning technologies so that it can require an industry to stop emitting pollutants. Without the available technology, government could not reasonably make such a demand. The Executive Branch might even need scientific evidence justifying regulatory activities before the Congress. In sum, scientists are needed by the Executive Branch because it becomes a combatant in the political process and therefore desires as formidable and defensible a position as possible.

THE REGULATIVE ARENA

Regulative policies develop in a relatively unstable relationship between the government and private sector. Regulation itself alters this relationship, since per se it involves a shift in government's role from laissez-faire to active combatant status. Government becomes a third player in previously private and two-person games, or it becomes a referee who changes the rules and awards points. With such an active role, government, including its Executive Branch, needs science and scientists to prove its case and justify its actions, generate ammunition for its arguments, advise on the consequences of its actions, generate acceptable solutions to problems or negotiate compromises, provide technological means to compliance with its regulation or adjustment to its deprivation, and define the existing situation the government might want to change.

Regulative games are zero sum, whether one considers the government's role or its choices. Its proper role in regulation or not regulating is a structural issue, while the choices it makes when regulating are allocative issues.[1] Regulative activities could also be seen as an integrated decision system variously rewarding, depriving, and choosing among several fragmented demands.[2] Government can act in an integrated fashion because its interests (votes, popularity, prestige) as a political unit would be advanced by regulatory activities. It also can act successfully because the objects of regulation are divided against one another, and government steps in to resolve those divisions.

Real regulation, it must be remembered, involves real choices, deprivations, or prohibitions. They must be tangible rather than symbolic choices, deprivations, and rewards. Further, as Murray Edelman would argue, real regulation is often not what is generally considered "regulation." Often what passes for "regulation" is a symbolic mask for distribution, self-regulation, or an agency or commission serving its clientele.[3] When a whole industry or private interest becomes aroused at the agency or commission charged with regulating that industry or interest, then that agency or commission is probably attempting real regulation.

Either wholly or in part, several policies could be considered "regulative." Conservation, transportation safety, trade and balance of payments, pollution, and antitrust policy are examples. However, securities and exchange policy, communications policy, organ transplant policy, water resources policy, population policy, and consumer or consumption policy might also be considered "regulative." In fact, the whole area of legal punishment of crime is "regulative." Such policy types are not wholly "regulative," but significant segments surely have that impact.

Scientists, as explained, are needed within the Executive Branch when it engages in regulative activities. Indeed, they are probably more useful to executive *agencies* engaged in "regulation" than to older independent regulatory *commissions*. Many independent regulatory commissions have been captured through the appointment process and long personal associations between regulator and regulated, whereas regulation by agencies is newer and more insulated by youth and by civil service procedures. Thus agencies might be more likely to want to "regulate," while commissions tend to distribute benefits to the "clientele" that captured them.

But scientists do not participate as policy-makers in regulative policy. They develop technologies (applied sciences or engineering), advise agencies, make authoritative and official recommendations, and define problems, solutions, and conditions. But they do not prescribe or administer policy. Economists are particularly involved, since the subject matter of so much regulation is economic as well as technological. They have access to the Executive Branch on agency levels and may often use the media and public opinion as a base (cf. conservation, pollution, trade). Policy, because regulation is so often politically sensitive, is made by politicians, lawyers, or administrators. But policy-makers do need science, if only to convince Congress, the courts, or the public that regulation is justified.

Often scientists define the problem in the media long before the government even considers regulation designed to correct it. Indeed, scientists' initiative has often led to Congressional initiative in setting up agencies with "regulatory" powers. Their "situation-defining" role should not be underestimated. But policy-makers, either Congressmen or executive leadership, cannot be indifferent to scientists if scientists are to have influence. On the contrary, both must agree with the thrust of scientists' evidence and the policy implications of these scientists' participation in policy making. Congress, in the regulative arena, is a significant point of access for scientists—and doubtless augments their influence.

Most policy perceived as having a "regulative" impact either deprives or chooses among private enterprises or the economic interests of producers. But behind most government efforts to regulate, there must be some notion of a general interest that regulation should promote or protect. Therefore, when regulation's impact is perceived as general (cf. requirement of exhaust emission devices on pre-1965 automobiles) rather than specific (cf. actions against one industry or producer), general support may decline. In one case, the general public is the benefactor of regulation. In the other, the general public could well see itself as being harmed by regulation (cf. requirement of emission control devices), even though the regulation might objectively be in the general interest. This is important because scientists' participation and influence rise and fall with the extent and success of the Executive Branch's "regulative" activities.

Scientists in a Regulative Arena: Transportation Safety Policy

While transportation policy may be primarily "distributive," transportation safety has increasingly become an issue for public debate and governmental action. But the Executive Branch's activities, backed with Congressional blessings, are seen by producers of transportation devices as "regulative." Therefore, the Executive Branch must continually contend with firms' hostility to government regulations—the auto manufacturers' distaste for auto safety requirements being an excellent example. Facing this hostility and its political power, the Executive Branch must have feasible and justifiable guidelines it imposes on manufacturers. Furthermore, it must make its case for the regulations (cf. mechanical causes of automobile accidents).

Rising concern has also enveloped the government's past efforts at "regulating" safety on common carriers such as buses, railroads, and airlines. It has been argued that government has not really "regulated" carrier behavior in these matters but deferred to industries self-imposed safety measures. Even the automobile tire industry has felt the rising concern, though it has not been a traditional object of regulation.[4]

Whereas pressures for developing new and better forms of transportation (distribution) came from political and scientific sources, the initiative for auto safety has come from scientists and Ralph Nader, a lawyer. However, despite the primitive quality of the science developed on safety and the low level of popular concern, "Nader's Raiders" were able to persuade the government to act. But given the fact that a reluctant executive with firm Congressional support decided to take "regulative" action, it needed scientific capabilities—statistics on accidents, sources of accidents, accident prevention technologies, highway safety campaigns, driver behavior and psychology, accident-reducing highway designs. Some of this "science" had been developed, but under industry sponsorship. Federally sponsored research was begun and used to shape guidelines and regulations. But scientists were needed, despite the primitive state of their science. No existing expertise or administrative experience could fill the void.

The relevant scientists came from the applied and engineering sciences. Their moderate influence over executive policy positions and activities stems from the executive's need for them. Policy makers have no other source of information, and they need scientific justification and technologies if they are to move against private interests in this field.

The public remains relatively unconcerned, though political leaders in attempting to "regulate" manufacturers assume, and probably quite correctly, that their action has a general payoff in support. But the political leaders have acted against a specialized interest and not against the general public, itself probably the major cause of automobile accidents. The public accepts traffic laws in general but may become aroused at specific laws. Suppose a state, in the interest of safety, began to require use of seat belts and enforced that law with frequent roadside checks. We could not expect general support to increase or remain constant. This can only occur when the ox being gored by "regulation" is another and specialized interest, e.g., auto manufacturers. Thus scientists' influence relates only to the regulation of specific rather than general interests.

Scientists have participated in policy-making as "outsiders." They tend not to hold administrative or formal policy-making positions but work through (1) efforts to mold public opinion, (2) testimony before Congress, and (3) advisory status in executive agencies charged with "regulating" safety. But even in the Congressional activities, their lay spokesman, Ralph Nader, has carried the day, building his arguments in part on the scientists' science.

Scientists have been divided over the political ends of their efforts; e.g., should government regulate safety. This division stems from the fact that many scientists, primarily the minimally innovative and nonactivists, are employed or under contract by vehicle manufacturers. These political and occupational differences have led to scientific conflict over data and findings. Yet, given the government's decision to act, it accepted the findings which supported its position. Furthermore, the government decided to develop its own scientific capability, realizing that the existing knowledge had been developed in part with manufacturers' funds in laboratories at Wayne State, UCLA, and Cornell. Then scientists, freed from the manufacturers' largess, would do research and publicly advocate safety measures.

Scientists, however, have been less a factor in transportation safety policy than Ralph Nader. In fact, without Nader's muckraking, popularizing, and catalytic role, government probably would not have been prodded into an active regulatory effort. Without Nader, safety regulation might be even more bounded and constrained than it is. Scientists would be even more hamstrung by a public obsessed with in-flight movies and horsepower rather than with safety, and corporations would be more concerned with design and cost than with accident prevention. The roles of Senators Abraham Ribicoff (D., Conn.) and Gaylord Nelson (D., Wisc.) cannot also be overlooked since they provided legislative interest and initiative. Similar essential roles by Nader and Congress surround efforts for airline safety, tire performance standards, and pipeline safety—fields where manufacturers are equally offended by proposals for regulation.

With legislation on the books, the crucial stage in policy making becomes the Executive Branch, with its invocation and application of guidelines. Manufacturers have political power, and though "regulating" activities and agencies are new, the industries' interests have already made a dent. Government can never impose "intolerable" burdens on manufacturers, and "intolerable" is often defined by the industry itself. Thus erosion of scientists' efforts begins immediately.

Regulation is a politically charged activity for policy-makers and administrators. The task of imposing regulations on manufacturers and users is a thankless job. Regulators need science, but they cannot politically afford to let it dominate decision making. There is a limit to the impositions administrators may make on manufacturers. There is an even tighter limit on the impositions administrators may make on users (cf. citizens wearing seat belts). Indeed, here as in many other policy situations, manufacturers and citizens want impersonal, technological, and quick solutions to problems. Manufacturers want the solutions to be cheap. Users want them not to change styles or patterns of living (cf. seat belts). These are powerful limitations, and they reduce scientists' influence and the extent of government's "regulative" activities.

Government, through its Executive Branch, has taken a "regulative" position from 1960 through 1968. This "regulative" impact was not extensively or fundamentally perceived either by executive policy-makers or by private interests early in the postwar period (1945–1960). Indeed, this original laissez-faire attitude lingers in the matter of transportation safety, but its days are numbered. And because the Executive Branch is increasingly and actively intervening among private interests and public needs, scientists' influence has increased.

Scientists in a Regulative Arena: Pollution Policy

Policy-making regarding man's environment—culminating in Congress' passage of the National Environmental Policy Act of 1970 and its creation of the Council on Environmental Quality in the Executive Office of the President—has often been closely related to scientists and the sciences. This has been particularly true in respect to one facet of policy's concern for the environment: pollution policy-making. Indeed, scientists (predominantly physical scientists, biologists, and ecologists) built the environment "movement" during 1969–1970 using the media and dragging government and political leaders along as "Johnnies-Come-Lately."

But fervor, rhetoric, and fad must not be confused with significant shifts in public policy. Pollution remains at best a matter for regulation and at worst a fad for conversation. Further, should, as Walt Kelly's Pogo, we meet "the enemy" and find "he is us," then it will be far more difficult to regulate ourselves than our industries. And, despite the cries of conservationists, for most Americans the problem of pollution has not become a matter of *communal security*. All of this inevitably must impose a limit on scientists' relevance and influence.

Governmental efforts at regulating pollution may involve air, water, noise, and land pollutants, all affecting man's physical environment. Scientists' moderate influence regarding pollution control activities in the Executive Branch has stemmed from (1) the government's intent to "regulate" private interests' behavior regarding pollution and (2) the scientists' efforts in defining and raising the problem of pollution.[5]

If government is to "regulate," as has been argued, then it needs scientists to define situations, develop arguments and justifications for regulation, and design technologies making compliance feasible. But here scientists' participation in policy making began earlier, when they pointed out the existence and effects of pollution and urged control of the environment. They successfully entered a political and technological vacuum. No expertise or awareness was present, but by 1960, when government began to actively regulate polluters and pollution, scientists' influence on policy making became quite noticeable. Thus this discussion deals with the recent 1960–1968 period of regulation.

The control of pollution means action on behalf of a general and long-term interest and against special, short-term interests. Wealth and economic power are affected. It costs an industry more if it is required not to pollute the air or rivers. But governmental action against pollutants can take two forms. Physical scientists and technologists may develop knowledge (technology) giving man and government power over nature and pollution. Techniques for cleaning up rivers and smoke control devices fall within this category, or economic and social scientists may develop knowledge about the behavior of polluters and the political situation that supports their behavior. Pollution can be controlled by changing technologies, behavior, or both. Technological solutions are easier, politically more acceptable, and quicker. But often technology is costly, and polluters are unwilling to bear the social costs of changing their behavior and life styles. Such economic and social costs may even inhibit elimination of backyard incinerators.

Furthermore, government will find that deprivations of private enterprises or industries that pollute the environment gain more public and voter support than deprivations of citizen polluters. An Executive Branch can take "regulative" action against a smoke-belching local industry but would well be quite timid in requiring *all* car owners to have exhaust emission control devices on their cars, new and old. The first action against industry does not substantially affect the life style or pocketbook of the average citizen. The second action does.

Thus "regulation" would seem to be limited to specific recipients that can be "regulated" one at a time or "divided and conquered." Regulative activities of policy-makers, as has been said, are politically based, and it would be impolitic to offend the general citizenry. Indeed, citizens apparently will applaud, though not initiate, regulative actions designed to protect their health or environment as long as that action does not cost them anything. We apparently will support protection of our health until protection requires changing our life styles, or until costs are directly imposed. Thus we resist emission control devices on our 1955-model "second" car, just as we resist pleas that we stop smoking. But if the government can clamp down on belching industrial smokestacks, we shout approval—just as when the government removes Thalidomide from the druggist's shelves.

Scientists from the engineering sciences, chemistry, biology, and economics have been active in pollution policy formation through the media and public, Congress, executive agencies, and the President's Science Advisory Committee. However, they have only begun to penetrate the core of the policy-making process, decision-making positions, or administrative posts. But, as scientists, they remain very influential advisers and "outsiders," particularly in diagnosing the problem and generating solutions. They led public opinion and saw first success when Congress seized the initiative on pollution control.

Their influence has come from their skills and Congress's strong efforts for pollution control. Indeed, many scientists, particularly those working for the Public Health Service (PHS), have been reluctant to become involved in administrative or prescriptive activities. The PHS's rejection of enforcement powers provided for itself in the Clean Air

Act of 1963 is a case in point.[6] Congress, however, prevailed and imposed enforcement powers on the Executive Branch. Its actions were backed by the American Medical Association, non-PHS scientists, and a willing public opinion that would at least tolerate if not applaud "regulation" of pollution.

Regulative action against industrial polluters can generate favorable public attitudes. Just as polluters have their political allies and base, so too have the antipollution forces. The polluters lost on the legislative battleground, and now their efforts shift to the Executive Branch charged with enforcement and with regulation. Here are new agencies, armed with scientific evidence and technologies and difficult to "capture" because they are new and scientifically sophisticated.

Private enterprise has vigorously opposed mandatory pollution controls, though major anticontrol groups (Manufacturing Chemists Association, National Coal Association, American Iron and Steel Institute, chemical industry, state and local governments, and auto manufacturers) did not actively coalesce against the Clean Air Bill of 1963, which passed with scientist and "urban-coalition" support. Some standard and inexpensive measures prove acceptable to most industries, and opposition may vary with the "reasonableness" of the regulative demands. Polluters' demands that regulation be "reasonable" should be appreciated, however, for a severe reaction awaits any Congress or agency that sets standards too high.

So, to some extent, consensus politics and compromise surround the application of pollution legislation, just as in transportation safety and health policy matters. Compared with most scientists' expectations, executive agencies' regulative actions are probably rather timid and undemanding. Consider that the first federal action on water pollution was taken in 1948 (Water Pollution Control Act). Even now conservation, scientific, and antipollution interests must goad administrators into raising standards and criticize them for falling under the influence of polluters and states wanting lower standards.[7]

Additional barriers inhibit influence for scientists desiring "regulation" of polluting activities. Public support for governmental action is hampered by a general feeling that skies and waters are free goods, that polluters upwind and upstream do not suffer when their pollutants drift downwind and downstream, that the private costs to polluters are greater than increased public costs of pollution,[8] and that increased industry costs may put local workers out of work due to higher equipment costs or industry relocating elsewhere.

The sciences dealing with environmental pollution have become fairly sophisticated, but conflicts still occur, including an ongoing debate on the effects of pollution on health.

Major conflicts of values and political views have often developed among scientists because so many scientists studying pollution and pollution control have been employees or consultants of industries. Their recommendations and science have reflected that association, just as conservation-oriented scientists reflect their own values and political commitments.

Further concern is expressed by ecologists and biologists that a pro-engineering bias in the National Academy of Engineering and the National Academy of Sciences panels, both formal advisers to the federal government, might bring a dominant pro-industry, progress-chained, solely economic approach to the pollution policy-making process.[9] Still, political forces and winds have blown toward "regulative" activities, and scientists have contributed with concern over pollution, analysis of its extent and effects, and technologies for its control. But the rather significant limits to their expectations and influence have been made equally apparent.[10]

Scientists in a Regulative Arena: Trade and Balance of Payments Policy

Government policies regarding trade and balance of payments matters have traditionally been "distributive" in perceived impact and have formed through logrolling or porkbarrel processes.[11] Elmer Schattschneider's classic study, *Politics, Pressures and the Tariff*, illustrates well the situation which prevailed from the late 1920s and even into the 1950s.[12] However, the 1960s witnessed a shift in the perceived impact of trade and balance-of-payments policy, away from the "distributive" manner in which government granted industries tariff requests and the "laissez-faire" manner in which government treated forces affecting the balance of payments.

Government's new posture is "regulative," denying and granting tariff protection in terms of the general and national interest rather than special industry needs. The study by Bauer, Pool, and Dexter of the Kennedy Trade Expansion Act and actions leading to its passage reveals this new post-1960 process. President Kennedy successfully redefined the issue in terms of national interests over special interests, and economists' arguments suddenly came to fruition and had relevance to policy.[13]

The Executive Branch, cleverly engineering Congressional support, began to regulate industries' benefits from tariffs, *selectively denying and granting* values such as wealth and respect which government once distributed on request. Indeed, the move toward lower tariffs and free trade can be construed as a clear regulative choice by government, acting in favor of consumers whose costs will be lower without protection because cheaper foreign goods could then enter the United States at competitive prices.

But even though government would act in favor of consumers and against producers, general support for government's position on "free trade" remains low, probably because it is such a complex, esoteric matter in which benefits are generally and indirectly dispensed to consumers.

Trade and balance of payments, despite the shift from "distributive" to "regulative" character, remain essentially political and diplomatic matters. The Kennedy Administration's efforts for "free trade" can be seen as a victory for consumers but also as a victory for a politically relevant group in American politics. Many large industries, facing no threatening foreign competition, and standing to benefit from lower foreign tariffs, supported the Trade Expansion Act. Industries and interests whose market was the world and not just the United States benefited from the action. These interests have close ties in most Administrations and Washington generally. As such "cosmopolitan" firms have emerged in the postwar period, so have those industries moved away from their "protectionist" positions.

The political character of trade policy is further demonstrated by a need for a few select exemptions (cotton, textiles, etc.) to the lower tariffs in the Trade Expansion Act of 1963. Thus the Executive Branch's decision to take "regulative" action in the general interest rather than to defer to special demands for "distribution" is itself a political decision built on the support of a variety of interests and opinions.

Trade and balance-of-payments policy making remains the province of the nonscientist. Economists' moderate influence has stemmed from long-term efforts at creating a "free trade" climate in the United States and from development of background information and analysis that support a general lowering of tariff barriers. But the immediate policy-making circles use economists only as advisers, even though many policy-makers, citizens, and corporation executives have been influenced by economists and the "free trade" climate they have created. Then, too, when the Kennedy Administration decided to frame the issue and legislation in the general interest, economists' calculations of the resultant increased trade and imports became important. Certainly, "regulative" activity demands and welcomes economists' participation and advice more than the "distributive" activities. The need for advice, justifications, and calculations is evident when the Executive Branch begins to make choices, even though the general thrust of the new policies is politically based.

Bauer, Pool, and Dexter's analysis, *American Business and Public Policy*, documents only one economist in inner policy-making circles. Gabriel Hauge on President Eisenhower's staff was directly involved in policy formation and is the exception to the rule. Furthermore, Congressmen indicated only minimal familiarity with the literature on trade economics. Domestic political and economic pressures rather than economists' arguments were partially successful in influencing economist-legislators, such as Senator Paul Douglas (D., Ill.). The Council of Economic Advisers' role was minimal.

It was, however, the general "climate" of support created by economists that paved the way. That contribution, coupled with the Executive Branch's need for advice on specific tariff levels and its decision to "regulate" trade matters actively, has meant significant economist influence. Indeed, the climate economists have in part created makes it unnecessary for them to participate actively in policy making, especially when rising economic interests (consumers and "cosmopolitan" industries with worldwide markets) benefit from the policies economists have supported.

Economists have been more influential on balance-of-payments matters, primarily because they less directly affect special interests and because they are seen as less "political" and "distributive" than contributions to trade policy making. "Regulation" to cope with balance-of-payments problems takes the form of cuts in overseas spending, fiscal policy, and monetary policy. These matters already fall within the purview of the Council of Economic Advisers and Federal Reserve Board, and thus scientific (economic) inputs are more readily and formally built into the policy-making process. Finally, the scope of interests and individuals involved in balance-of-payments policy making are considerably less than the situation for trade policy making. The narrower confines, coupled with less "politics" and a clearer notion of the national interest makes the policy process more hospitable to economists.

Scientists in a Regulative Arena: Conservation Policy

Conservation policy stems from governmental activities affecting the preservation or selective consumption of natural resources, such as ores, forests, and natural scenery. The object of preservation or selective consumption is the assurance of supplies for such future needs as manufacturing, fuels, and recreation. Thus such techniques and subjects as erosion control, forestation, ore extraction, plant genetics, and federal protectorates over scenic or wildlife areas are involved. The government, through its Executive Branch, may allow foresters or miners to exploit resources under a "laissez-faire" umbrella, encourage resource exploitation, prevent or limit resource exploitation, or exploit resources itself. Conservation activities are located in the Departments of Interior and Agriculture, and range from the establishment of national monuments to management of forests to authorization of dams.

Scientists have been moderately influential in conservation policy making.[14] Their participation and conservation activities themselves have not fundamentally changed in the 1945–1968 period. With some exceptions, government activities continue to be "regulative" in perceived impact. This pattern was first established with the "conservation movement" early in the 1900s, and it continues, with government increasingly moving away from laissez-faire or distributive policies and toward regulation. The values involved range from well-being and respect to wealth. The benefactors of governmental "regulation" are future generations. Though the present general public may pay higher costs for products now, it also receives its own immediate benefits in recreational facilities.

Basically, however, regulation for conservation purposes deprives the present in favor of the future. Various conservation measures selectively benefit and deprive current interests. Resource-exploiting and resource-consuming industries achieve wealth as a value, while conservation organizations strive for such values as respect for their style of life, recreational opportunities, and a sense of well-being.

Conservation movements are permeated with scientists. The sciences of biology, botany, economics, genetics, and chemistry have relevance as scientific inputs for policy making, but conservation issues have aroused and involved far more scientists than their scientific contributions would indicate. Grant McConnell, writing about the early years of conservation movement, notes that

> Perhaps the most striking then was the number of natural scientists who lent their names and energies. They included geographers, geologists, botanists, and others of even more retiring professions.[15]

Apparently most scientists prefer and value conservation measures, *at least* for resources such as natural scenery and wildlife.

Scientists have not occupied conservation policy-making positions, though they have access at most points in the governing process. This influence has stemmed from opinion-making activities in the public media and conducting research on the substance of conservation policies, such as erosion control, deforestation, economic planning, and hybrid development. Scientists' influence and government "regulation" have been minimal when an industry exploits a resource at its own pace and does not disturb natural scenery or "conservation-sensitive" issues. Oil extraction and coal mining are examples. Normally, firms exploit resources for maximal profit, but occasionally their activities disturb other interests and conservationists. Strip mining performed in Kentucky, Ohio, West Virginia, and neighboring states has become a sensitive issue. Oil leakage from a well in the Santa Barbara Channel (California) recently led to activation of conservationists. It is on these occasions (construction of dams, timbering of national forests, commercial development of ski areas on federal recreational lands, etc.) that scientists become active and influential through conservation interests.

But when the sole issues are economic use of resources and "savings" for future generations, scientists have made almost no dent in a "laissez-faire" government-industry relationship. Only when the specter of "waste" and symbol of "conservation" are aroused do they have influence through the media and agencies.

"Conservation" is a powerful symbol evoking fundamental and significant support from Americans. Therefore, a President or Congress that stands with "conservation" against "exploitation" has wide support and stands to benefit politically. As a result, conservation policy is still based on political considerations more than on scientific evidence. In fact, if an agency calculates that it can allow timbering on federal lands by private interests, and *not* arouse too much conservationist protest, it may well do so for political reasons, irrespective of economic or resource studies.

Indeed, conservation's loudest voices have come from political figures who "invented" the movement and built careers on the issue—Gifford Pinchot and Theodore Roosevelt. Their appeal was based on benefits to the "public interest," future generations, and an avoidance of waste.[16] Since their time, conservation policy making has acquired the qualities of a developed policy arena. The Executive Branch has remained committed to conservation measures and derives general support from those activities. Scientists' influence has depended on this political commitment.

Conservation policy did not exist during the "era of plenty" in early American history. It only became a concern when scarcity cast its shadow. Science itself long suffered from this "abundance psychology" and was not aware of any possible need for man to conserve. Therefore, the conservation-related sciences had to create and await an awareness of conservation problems. But even now, with general awareness assisting their efforts, scientists and the executive agencies must work within the demands of business, industry, lumber interests, timber owners, and organized grazing interests. These demands prevent a full-blown conservation effort based on economic, physical, and resource sciences. What efforts have been made have been based on sympathetic executive leadership and favorable public opinion, itself skilfully shaped by conservationists and politicians.

Scientists generally agree on the desirability of conservation measures, but they often divide over scientific issues. Also political and ideological considerations may lead to suppression of scientific evidence. Ashley Schiff's *Fire and Water: Scientific Heresy in the Forest Service* illustrates this problem for scientists.[17]

Traditionally, the Forest Service maintained a position against fire or burning. Service scientists shared this dogmatic belief along with their political superiors. However, beginning in 1907 scientists working outside the bureaucratic structure began to chip away at the reigning ideology, which even then was a quasi-religion and *idée fixe* promulgated with evangelistic fervor.[18] These scientists' efforts spanned the 40 years until 1947, when Smokey the Bear became a more realistic and less dogmatic ideology.

Two factors rendered the Forest Service impervious to scientifically initiated discovery and change:

1. Older scientific theories persisted and conditioned the values, perspectives, and goals of administrators and scientists doing "in-house" research. These theories included the notion of a stable climax forest, which needed no human intervention to save the forest in its present form; the Soil Conservation Service's notion of a static level of soil maturity, which prevented work on productivity and soil depletion; and a general preoccupation with static rather than dynamic concepts of nature.

2. Organizations, the Forest Service included, tend to develop or sustain scientific arguments that buttress their intentions and programs, and refute others' efforts to change those programs.

Therefore, the Forest Service resisted H. H. Chapman's early findings that long-leaf pine needed fire to survive and reproduce and that a "no-burn" policy meant an incursion of slash pine on long-leaf stands.

Similar resistance was encountered in dogmas about "no-cutting" and erosion in forests. Findings that selective deforestation could improve drainage and run-off after rainfall, thus preventing flooding, were resisted for many years. These conditions have some remnants in the 1960s, even though the general climate surrounding "in-house" scientific research has become freer and more tolerant. But, through a historical flashback, the baneful effects of embedded organizational ideologies have been revealed, something Schiff expresses when he notes that:

> Traditional theories of organization ontogeny suggest that, as barnacle encrustation progressively impairs ship performance, so may administrative structures lose their responsibleness through time. Thus, institutional inflexibility is usually deemed as infirmity of advancing years. While this pathological diagnosis has considerable merit, it ignores the effect of militancy on agency behavior.[19]

The extreme conditions Schiff so well documents do not measurably exist over this 1945–1968 period, but they do illustrate the continuing political character of conservation policy making.

Scientists' influence now depends not on agency "blindness" to science but on the "regulative" balance between government and private interests. They influence executive agencies insofar as those agencies attempt to "regulate" special interests. But scientists' most significant activities continue to occur in the public realm, often outside their normal scientific competence and fields, and promote "conservation" much as the postwar atomic scientists pursued "civilian control" and "international controls" in the public arena and outside their immediate scientific specialization. But then, neither "conservation-oriented" or "weapons control-oriented" scientists could make much measurable impact on policy if they had no reputation as productive scientists and if no body of scientific knowledge on the policy existed.

Scientists in a Regulative Arena: Antitrust Policy

Government's "regulative" activities in the antitrust field involve the allocation of such values as wealth and power in the economic sphere. In particular, this means antitrust actions of the Department of Justice (Sherman Antitrust Act, mergers, market control) and competition-promoting actions of the Federal Trade Commission (Clayton Act, pricing, anticompetitive practices). The period from 1960 to 1968 has been selected for analysis because it encompasses a new sophistication in the use of the economic sciences vis-à-vis antitrust policy. Generally, however, governmental activities in the antitrust field have always been "regulative" in impact, at least as perceived by private enterprise.

Antitrust actions have been *political* choices since first initiated. The premise that government would attempt to control corporate and market structures stemmed from political criteria. Economists since 1900 have increasingly refined their potential contributions to antitrust policy making and concurrently influenced the *means and criteria* through which the Executive would pursue its politically based objective.[20]

Still, economic criteria are the lesser component of antitrust policy making. It is a politician's and lawyer's field. Economists diagnose the problems created by monopoly or anticompetitive practices, set forth alternative courses of action, and provide economic rationale for existing policy, which itself would be politically based and motivated.

Economists indirectly were involved in early antitrust actions and provided economic statistics on firms' market shares and size of operation. However, they gradually evolved new definitions of market structures, "competition" or "monopoly"—definitions that would eventually wend their way into the antitrust lawyer's brief and from there into the courts.

Clearly, legal and economic notions of "monopoly" and "competition" have become increasingly sophisticated. No longer could a Department of Justice lawyer find such obvious violations as those of the robber baron and his overt flaunting of "morality," predatory practices, and blatant collusion. Business has matured and violations of antitrust laws must be more subtle and "technical" rather than "immoral." Thus the Department of Justice needs economic evidence and understanding, if only to buttress and provide rationale for its politically determined antitrust activities. But because antitrust actions evoke and occur within a political context, economists' influence is limited by the reigning political intentions of a particular historical period or Administration.

Economists themselves do not occupy administrative or policy-making positions. They are advisers and an academic community that gives leadership to refined notions of "monopoly" and "competition." They have led the times in anticipating shifts in antitrust criteria, but beyond setting the stage, they have been constrained by political and legal factors dominating the relationship between the Department of Justice, the courts, public, and corporate community.

The role of economists as participants in antitrust policy making did not fully mature until Thurman Arnold's era at the Department of Justice, when economic evidence and definitions became major factors in decisions, policy, and court cases. Originally, trustbusters had acted only against "sin" and not "size," but during Arnold's tenure, they came to consider size itself as a criterion of monopoly. This was true under antitrust chief Arnold when:

> The employment of a number of economists enabled the Division to focus more on market data evidence and less on allegations of predatory practices, thereby paving the way for greater judicial acceptance of economic criteria of monopoly or restraint of trade. A team of attorneys and economists could now be detailed to make a comprehensive investigation of an entire industry, or a type of market control wherever found, concentrating on the focal points of price and output

determination and laying the groundwork for formal attacks through the courts.[21]

The use of economic criteria begun under Arnold developed more sophistication after 1960. The new criteria and rationale for antitrust activities were found in the notion of "workable competition." Size and sin were no longer major factors. Indeed, the intent to monopolize or eliminate price competition did not have to be present.

Some industries would be less aroused by this more sophisticated policy, inasmuch as it recognized their own interests. And the new policy and definition appreciated the status quo. Profit was recognized as a valuable and appropriate objective. *Workable competition* involved:

1. A fairly large number of buyers and sellers, each with no dominant share of the market.
2. Absence of collusion or conscious parallelism of behavior.
3. Free entry of new firms or buyers.
4. Credit and recognition for business performance (i.e., lower prices, efficiency, product improvement).
5. Approval from the Department of Justice prior to mergers.
6. "Bigness" not being per se "badness."
7. Need for increased competition and decentralization but no unreasonable demand for either of these.

This represents increased sophistication in government's understanding of the economic system and a more "reasonable" approach to balancing competition and production. But it has meant that industry is both freer (bigness is no violation of the law) and constrained (conscious parallel movement of prices as a violation of the law).

Generally, business seems less disturbed with the new antitrust philosophy and norms. Perhaps there is emerging a new coincidence of interest and attitude among economists, judges, firms, antitrust lawyers, consumers, labor, and the "public interest." Modern firms are not as vulnerable to the antitrust zealots because their aggressive behavior is less blatant and within the "moral" limits of competition. Furthermore, antitrust notions have not been extended to labor unions and agriculture, in part because of political realities and in part because economists have not sufficiently extended their analysis to those sectors.[22] Modern firms also tend to accept the newer antitrust criteria because they promise some preservation of the status quo and continued competition, both of which serve the interests of each firm. This new "harmony" results as much from a modification of economists' positions and more "reasonable" Department of Justice criteria as from the changed nature and behavior of the corporate sector.

But antitrust policy making still does not qualify as a conflict-free arena. Firms continue to see the Department of Justice's Antitrust Division as an "uncongenial" agency—not one of their closest friends. Most firms could do without the Department. For established firms in the long run, the Antitrust Division and Federal Trade Commission are frequent but insignificant "pains-in-the-neck" or brush-fires to be fought and tolerated.

However, occasional situations arise when private enterprise meets government head-on. Such cases would be the jail sentences imposed on General Electric executives for price fixing and current concern over the need for antitrust action against "conglomerates" and "one-bank holding companies."[23] When these confrontations occur, government must have a fully developed court case, and this means it must use economists' contributions. This is necessary, it is argued, because government, when "regulating" the behavior of special interests vis-à-vis the consumer sector or other special interests, needs scientific justification and evidence to prosecute successfully such a risky political course. To maintain its case before a supporting public, other firms, and the courts, government needs the sophistication economists can lend to its arguments, regulations, and actions.

Notes

1. Robert H. Salisbury and John P. Heinz, "A Theory of Policy Analysis and Some Preliminary Applications," *Policy Analysis in Political Science,* ed. by Ira Sharkansky (Chicago: Markham Publishing Co., 1970), pp. 39–60.

2. *Ibid.,* p. 3.

3. Murray Edelman, *The Symbolic Uses of Politics* (Urbana: University of Illinois Press, 1967), pp. 23–29, 36.

Louis Kohlmeier's *The Regulators* documents symbolic regulation and regulators' failure to promote the public or consumer interest. His work shows how regulatory agencies and commissions cannot fulfill both the promoter and regulator role. Numerous examples of meaningful "regulation" and "symbolic" regulation are set forth. See Kohlmeier, *The Regulators: Watchdog Agencies and the Public Interest* (New York: Harper and Row, 1969). Kohlmeier discusses regulation in general and focuses specifically on transportation, labor, banking and investment, energy, television, and trade, showing how economic criteria, technology, and political demands do not mix.

4. Some sources on transportation safety policy and scientists are E. R. Piore and R. N. Kreidler, "Recent Developments in the Relationship of Government to Science," *Annals of the American Academy of Political and Social Science,* CCCXXVII (January, 1960), pp. 10–18; Robert Burkhardt, *The Federal Aviation Administration* (New York: Frederick A. Praeger, 1967); Grant S. McClellan (ed.), *Safety on the Road* (New York: H. W. Wilson Company, 1966); U.S. Congress, House, Research and Technical Programs Subcommittee of the Committee on Government Operations, *The Use of Social Research in Federal Domestic Programs,* 90th Cong., 1st Sess., April, 1967, Volumes I–IV.

5. Some sources on pollution policy making and scientists are Marian E. Ridgeway, "The National Water Pollution Control Effort," *Quar-*

terly Review of Economics and Business, III : 1 (Spring, 1963), pp. 51–63; Randall B. Ripley, "Congress Supports Clean Air, 1963," *Congress and Urban Problems,* ed. by Frederick Cleaveland (Washington: Brookings Institution, 1969); Marshall I. Goldman, *Controlling Pollution* (Englewood Cliffs, New Jersey: Prentice Hall, 1967); Donald E. Carr, *Breath of Life* (New York: W. W. Norton and Co., 1965); Gladwin Hill, "The Politics of Air Pollution," *Arizona Law Review,* 10 : 1 (Summer, 1968), pp. 37–47; and J. Clarence Davies, *The Politics of Pollution* (New York: Pegasus, 1970).

6. Ripley, *op. cit.* Gladwin Hill also notes that the Air Quality Act of 1967, though it added no more federal controls or authority, did increase research funds.

7. *Science,* 160 (April 5, 1968), p. 49.

8. Mancur Olson, Jr., *The Logic of Collective Action* (Cambridge: Harvard University Press, 1965). Olson argues that individuals polluting the environment would have to incur significant costs were they to curb their own pollution voluntarily. Society would, he admits, benefit from their action—total pollution would be decreased. But their costs (control devices) would exceed their own benefits, since much of the pollutants they emit would, without controls, end up in someone else's lungs, water, or yard. Thus pollution controls must be mandatory and will even then be resisted by the economically rational.

9. *Science,* 160 (January 19, 1968), pp. 287–289.

10. Personal conversations with Charles O. Jones have sharpened my understanding of the formation of pollution policies, but he is not responsible for my conclusions.

11. The basic studies of trade policy making are Raymond Bauer, Ithiel de Sola Pool, and Lewis A. Dexter, *American Business and Public Policy: The Politics of Foreign Trade* (New York: Atherton Press, 1964) and E. E. Schattschneider, *Politics, Pressures and the Tariff* (Englewood Cliffs, New Jersey: Prentice-Hall, 1935).

12. Schattschneider, *op. cit.*

13. Bauer, Pool and Dexter, *op. cit.,* pp. 24, 437.

14. Sources dealing with scientists and conservation policy include Luther Gulick, *American Forest Policy* (New York: Duell, Sloan and Pearce for Institute of Public Administration, 1951); Ashley L. Schiff, *Fire and Water: Scientific Heresy in the Forest Service* (Cambridge: Harvard University Press, 1962); Frank E. Smith, *The Politics of Conservation* (New York: Pantheon House and Random House, 1966), Ashley L. Schiff, "Innovation and Decision Making: The Conservation of Land Resources," *Administrative Science Quarterly,* II : 1 (June, 1966), pp. 1–32; Grant McConnell, "The Conservation Movement—Past and Present," *Western Political Quarterly,* VII (September, 1954),

pp. 463–478; and Dean E. Mann, "Politics and the New Conservation" (Unpublished paper, Earth Sciences Colloquium, University of Arizona, Tucson, December, 1969).

15. McConnell, *op. cit.*, p. 463.

16. *Ibid.*

17. Schiff, *Fire and Water*, and Schiff, "Innovation and Decision Making."

18. Schiff, *Fire and Water*, p. 115.

19. *Ibid.*, p. 165.

20. A primary source on antitrust policy making is Merle Fainsod, Lincoln Gordon and Joseph C. Palamountain, Jr., *Government and the American Economy* (New York: W. W. Norton and Company, Inc., 1959), Part 4, pp. 427–618.

21. *Ibid.*, p. 573.

22. Carl Kaysen, "Model-Makers and Decision-Makers: Economists and the Policy Process," *The Public Interest*, 12 (Summer, 1968), pp. 80–95. I would agree with Kaysen that the absence of "antitrust" action against unions and agriculture indicates reluctance of government to enter these fields, particularly as a "regulator." However, Kaysen depreciates the influence of economists for the same reasons. But economists have not sought free competition as a norm in all economic areas, and neither do they seek that "ideal" the corporate sector. Their influence must be judged partly in light of their intent to influence.

23. Antitrust action against conglomerates can partly be attributed to political pressures from existing nonconglomerate giants fearing the rise of new economic power to challenge theirs.

The Policy-making Process and High Levels of Scientist Influence

CHAPTER 6

The economic management, communal security, and entrepreneurial policy arenas harbor conditions conducive to a high level of scientist influence. Scientists' status as highly influential participants depends in part on two major conditions.

First, government may be an "entrepreneur," either producing a product (space exploration, atomic energy, public power) or heavily endowing a national effort (support of scientific development). But for scientists to be influential in an "entrepreneurial" situation, either scientific considerations must permeate the production process (space exploration) or scientists' own interests must be affected by government's entrepreneurship (science-policy support for research and development).

Second, government may be acting both in its own and the common interest,

NOTES TO CHAPTER 6 START ON PAGE 221

preferably providing a common sense of security. This "security" may involve freedom from economic recession (fiscal and monetary policy), disease (health policy), forces of nature (weather policy), or military threats to national security (weapons policy, deterrence and defense policy). The impact of government's activities, in these cases, is seen as a general or communal distribution of security as a value. But the security accrues to government and citizenry alike, for both are threatened and need security.

Therefore, when science and scientists are affected by policy making, especially the distribution of funds to scientific research, they are bound to be influential. When scientific considerations dominate a governmental "entrepreneurial" effort, the government must involve scientists in policy-making and administrative positions. When the security of citizen and government is commonly threatened, scientists are needed if the government is to survive politically and physically by producing these common benefits (freedom from economic disaster, disease, weather, and military threats).

Scientists occupy important policy-making positions, ranging from prescription and invocation to administration. They are policy-makers, because they are deemed highly competent scientists by political leaders and because they are needed. But the fact that scientists and their science are needed in such crucial fields as military security, economic management, and health has meant that vast amounts of funds have been granted to scientific research in these fields. Government has created "competence" because it needs any level of competence it can get and because it prefers a higher level of competence. Therefore, both competence and influence are shaped by political demands; i.e., government's need to provide communal security and government's desire to be an entrepreneur in scientifically laden fields.

The need for scientists stems from a high level of urgency in providing the demanded "communal security" benefit. No division or indecision plagues the Executive Branch's leadership. No hostile vested interest opposes scientists' efforts. In fact both private enterprise and the scientists' own vested interests are invariably benefited by scientists' participation in policy making. The general political climate supports and needs the scientists' contributions.

The sciences involved range from economics to physics, but social sciences are needed less than other sciences in these arenas. Comprehension of the "hard" sciences is deemed more difficult by policy-makers. The social sciences are "comprehensible" because their knowledge can be acquired by the policy-maker himself through common sense, intuition, and experience with people. But policy-makers confronting physicists and chemists meet more specialized and highly developed fields bearing no relation to the policy-makers' experience or frame of reference. I do not deny, however, that policy-makers may actually have more difficulty understanding people (social sciences) than things (physical sciences). But the important fact is how the policy-maker feels he comprehends a science.

Scientists are deemed so important to the objectives of governmental activities (security, successful government "entrepreneurship") that they have access ranging from the media to Congress and the Executive Branch's higher levels. The major point therefore remains, that scientists are highly influential because they make an essential contribution to a communal sense of security (health, weather, weapons, defense, fiscal and monetary policy), enable the government to produce a product (space policy), or see their own interests directly affected by government's distribution of funds (science policy).

THE ECONOMIC MANAGEMENT ARENA

Policy formed in the economic management arena attempts to control fluctuations in the nation's growth rate, unemployment levels, and price levels in deflationary and inflationary situations. In a larger sense, efforts at managing the economy are spurred by memories of the depression of the 1930s.

The common economic insecurities of the government and citizens demand economic stability. Government is deemed responsible for economic security and welfare, and Keynesian economics urges and enables government to assume such responsibility.

The demand that government manage the economy to achieve a level of economic security, coupled with similar demands for communal security from disease (health) and military threat (weapons, defense), has led America to become in part a *security state*. So highly do we value security in these matters, that government must develop as much science and use as many scientists as possible to make its actions more appropriate and effective. No government or citizenry is overly willing to take chances with its own economic, military, or personal security.

Fiscal and monetary policy has not always been shaped by a dominant concern for "security." Taxation, particularly the issue of progressive income taxation, was redistributive in its perceived impact. But now, the manipulations of the government's budget (fiscal policy, manipulating taxation and expenditure) are seen more to "distribute" economic security to all citizens. Disregarding the impact of *particular* expenditures and differential rates of taxation, a tempered pattern of growth and stability benefits most major groups in society, just as a military "umbrella" protects all groups in society. A citizen can no more exempt himself from protection against nuclear attack than he can deny he is a benefactor of government's fiscal policies and economic security.

John Kenneth Galbraith argues in *The New Industrial State* that firms employ a "technostructure" and economists to insure freedom from economic uncertainty. Uncertainty, and the insecurity that may follow from it, are alleviated by prediction and anticipatory policies, creation of new markets

The Economic Management Arena 171

and products, planned obsolescence, and government's activities in encouraging economic growth and adequate demand.[1]

Economic management is a matter of allocating benefits (economic security) to the society. The situation and policy are seen as nonzero sum. Government regulates the economy in order to distribute economic security. Various fiscal and monetary policy actions benefit different groups differently, but the larger purpose of "managing the economy" dominates the arena. Conflict between "haves" and "have-nots" or between consumers and producers are subordinate to the overall common objective—a depression- and instability-free economy.

Scientists, predominantly economists, have had a *moderately high* level of influence on economic management activities within the Executive Branch. This condition has persisted in a stable set of relationships since (1) the establishment of the Federal Reserve Board in the early 1900s and (2) the creation of the Council of Economic Advisers through the Full Employment Act of 1946. But government's responsibility for managing the economy has never been deemed so important as in the 1945–1968 period.

Economists have had institutionalized access to the President, Federal Reserve Board, and high levels of the bureaucracy. They have become policy-makers without administrative responsibilities. They are closely interwoven within the policy-making process because decision-makers need their expertise and because it is so important that governmental actions in managing the economy be successful.

The emphasis placed on the "distributive" and "regulative" impacts of policies in the economic management arena should not obscure the ease with which an issue may become "redistributive." Economic issues can easily become divisive when individuals come to believe that their particular costs and benefits outweigh any benefit accruing to the total community. In other words, the deprivation of a group by a particular tax, expenditure, or credit policy may lead that group to withdraw its support of an overall fiscal or monetary policy designed to insure communal economic security.

Government often combats or anticipates this possibility by stressing the disaggregation of the collective benefit. In this case, government would argue or insure that a policy deemed "redistributive" in impact really does benefit groups that feel they do not benefit. For instance, government would argue that increased spending or taxation enables not only economic stability (and security) but also increased markets for firms, reduced ghetto unrest, correction of "eyesores" through urban renewal, increased public facilities, and, even if policy carries a tinge of "redistribution," an appeal to individuals' sense of social equity. Government simply shows that measures promoting communal security concurrently promote various groups' own values and interests.

Scientists in an Economic Management Arena: Fiscal and Monetary Policy

Economists have had institutionalized positions as advisers and policy-makers in the Executive Branch since 1945, when postwar concern with economic stability generated pressure that led to the Employment Act of 1946. Since that time, Americans have been desirous that their government regulate the American economy to ensure (i.e., distribute) economic stability, maximum employment levels, growth, and a sense of common economic security. However, these expectations of government and government's use of economists to fulfill those citizen expectations did not become established until 1960. During the 1960–1968 period, economists have increasingly filled more effective advisory roles and have become integral parts of the policy-shaping process to such an extent that their influence can be described as moderately high.[2]

The unique character of scientists' involvement in fiscal and monetary policy-making within this economic manage-

The Economic Management Arena   173

ment arena lies in the fact that they are *economists*. And economics differs from the physical sciences so closely involved with the communal security arena (health, weapons, weather, deterrence, and defense policies). Still, most Americans' common economic security is a major value affected by economists' contributions in the economic management arena.

FISCAL POLICY-MAKING. Fiscal policy involves government's efforts to ensure price stability, full employment, and growth by manipulating its budget (taxation, expenditure, deficits, and surpluses) to affect the nation's economy. Economists participate in fiscal policy-making as either consultants or members of the Council of Economic Advisers.

The Council's early years (1946–1952) were filled with conflict and adjustment as it sought out a role, carving its own niche within a policy process that has been dominated by political considerations and political leaders. But no major internal disputes have racked the Council since that early conflict between members Leon Keyserling, Edwin Nourse, and John D. Clark. The Eisenhower Council under Arthur F. Burns was more successful in working with and guiding the President than Keyserling had been with President Truman. Truman apparently held measurable biases against scientific, including economic, expertise.

The Keyserling Council was successful, but its effect probably depended more on the demands of the times, which fortunately gave impetus to the very measures the Council advocated. The Council's expansionist ideology dovetailed philosophically with the Fair Deal, but was personally out-of-step with the more stability-oriented and conservative-oriented economics of Truman and a public that was not yet ready to accept the ultimate consequences of heeding economists' advice. Still, the expansionist policies Keyserling advocated fit nicely with the Korean mobilization and with the military development planned in a National Security Council report (NSC–68). Keyserling in fact oversaw preparation of the economic section of NSC–68.

Likewise, the Burns Council was successful largely because its arguments dovetailed with the prevailing winds and political goals of the Eisenhower Administration. Even so, this Council had to contend with the dominant role of Secretary of the Treasury George Humphrey in economic matters. Nevertheless, during the chairmanships of Burns and Saulnier (1952–1960), the Council's image measurably improved in the public eye and set the stage for a full blossoming of its participation in policy making during the Kennedy Administration.

The Council of Economic Advisers under Walter Heller developed new significance and status *within* the fiscal and monetary policy process. The educability and receptiveness of President Kennedy was a clear factor, despite the fact that the Council's arguments again had to merge with overall Administration positions. Kennedy apparently could comprehend the economists' arguments and engaged them in a productive policy-centered dialogue. The Heller Council "developed and gained acceptance for the economic philosophy upon which the (Kennedy) tax program was built." [3] Kennedy had a lightly held belief in budget balancing when he assumed office and, as Herbert Stein argues, changed his views because:

> He was not the first Keynesian President on Inauguration Day, but he was the first who was not a pre-Keynesian—the first who had passed the majority of his life in the post-Keynesian world where the old orthodoxy was giving way to the new.[4]

The lessons of John Maynard Keynes were driven home, 15 years after they had been nominally recognized in the Employment Act of 1946. Government would actively intervene in the economy to ensure economic growth and security, primarily by manipulating aggregate demand or consumption in the economic system. The Heller Council had the further advantage of a more mature science of economics, a renovated discipline invigorated by new methods, sophistication, and professionalism.

The development of more sophisticated quantitative economic models and econometrics has probably introduced a new asset for economists' influence through post-Heller Councils. Such new techniques would lend an additional aura of accuracy, and policy makers might be more likely to perceive economics as more "scientific." Consequently, more scientifically-based prediction and prescription join the economists' portfolio, alongside ideological and policy-oriented advice.

Post-Heller Councils under Gardner Ackley and Arthur Okun in the Johnson Administration continued to be influential as insiders in the policy-making activities of the Executive Branch.

The Council of Economic Advisers since 1960, despite its influence, has had no management or administrative functions. It has served as a source of ideas, proposals, information, and policy advocacy, while concurrently being closely integrated into policy-making circles. It has provided the economic rationale for the economic policies expounded by Administrations but, at the same time, has increasingly molded those policies.[3]

A President chooses his economic advisers, but because economics has changed in Keynesian directions, he cannot choose among economists with competing scientific viewpoints.[4] Speaking of Kennedy's choices, Herbert Stein says:

> If he [Kennedy] had chosen six American economists at random, the odds were high that he would have obtained five with the ideas on fiscal policy that his advisers actually had, because those ideas were shared by almost all economists in 1960. . . . Kennedy did not choose his advisers to advocate and practice a particular brand of fiscal policy upon which he had already determined. He chose them as representative of the economics of his time, and having done that, he exposed his policy to influence by the economics of that time.[5]

High Levels of Scientist Influence

There is little question that the economists shaped each President's specific response to the times. Indeed, since 1946 the Council has had some "liberalizing" effect on the incumbent President. It has moved Presidents to activity rather than inactivity on economic issues (cf. Keyserling's expansionist philosophy, Burns's preference for countercyclical actions by government, and Heller's notions on deficit spending).[6] This positive role appropriately complements the Bureau of the Budget's traditional negative and "nay-saying" function. Indeed, the Council's "successes have derived from getting things started or changed, and its failures from being unable to prevent things from being undone or delayed or stopped."[7]

Still, the policy-making role should not be overemphasized at the expense of the advisory role. The Council has been more involved and influential at some times than at others, and the trend seems to indicate increased influence; but full penetration of policy-making circles may always lag. Edward Flash, Jr., concurs, noting that:

> The demand for the Council's services is also affected by the nature of their application. The demand for information and comment, for ammunition and rationalization, for drafting and verbalizing is generally greater (that is, less elastic) than the demand for direct participation in the creative and ratifying steps of decision-making.[8]

Fiscal policy must still be made within a political environment. Citizens prefer deficits to tax increases, spending cuts to tax increases, and surpluses to deficits, depending on the specific situation. Thus an economist recommending tax cuts will be more influential than an economist pressing for increased spending. This is precisely what happened early in the Kennedy Administration.

Tax increases or tax surcharges will be perceived by significant political groups and the general citizenry as having "regulative" impacts. And, as has been earlier noted, regulation of a general interest is more difficult than regulation of a special interest. At least some segments of the population are neutral or supportive when someone else's "ox is being gored."

But because tax increases may be economically appropriate and because political support is necessary to Presidents, the Council, particularly through its chairman, has taken on a new role as leader of public opinion, economic spokesman for the Administration, catalyst, and public educator.

MONETARY POLICY-MAKING. Monetary policy, developed by the Federal Reserve Board, involves efforts to control demand and consumption in the economy by restricting the amount of available credit. Policy making by the Board is clearly a more stable, ordered, and tradition-bound process when compared with the Council of Economic Advisers' still-emerging role. Federal Reserve Board economists, however, play a more restricted "advisory" role well backstage in comparison with their CEA counterparts.

Still, some economists' influence over monetary policy has noticeably increased since the early 1950s, as they have begun to speak publicly and take on advisory status with a government and banking system plagued by rising complexities and difficulties in responding intelligently to the modern economy. By 1961, six of twelve Federal Reserve Bank presidents were professional economists. Gradually, the Board in Washington has come to consider an economics background as providing "better credentials than law or banking" [9] for new members. No Federal Reserve Bank directors in 1925 were from academic circles, while many now emerge from that background. Furthermore, one of President Johnson's appointments to the Board was a former Reserve economist, Andrew Brimmer.

Economists who have become policy-makers and Board members are relied upon more than the regular economists on the Board's staff. Noneconomists seem to turn to fellow policy-makers with economic backgrounds rather than staff experts. The staff is a source of statistical information and economic research, while economists who have become policy-makers and Board members are sought out for new ideas, such as theories on central bank policy and appropriate monetary actions. The "bills only" policy (open-market operations only with Treasury bills), tight money policies during the Eisenhower years, and opposition to selective instruments of control beyond use of margin requirements were such innovations stemming from these economists.[10]

Indeed, members with economics backgrounds have reached relatively equal status in policy-making with members possessing "traditional" backgrounds in banking and management, even though bankers often see themselves as their own economists with firsthand knowledge of the operations of the economy. Monetary policy-making is more likely to pit expert against expert, while fiscal policy-making confronts economists with politicians having much less experience in economic analysis.

Economists' participation, either with the Federal Reserve Board on monetary policy or the Council of Economic Advisers on fiscal policy, has developed significantly since 1960. These economists have successfully penetrated the policy-making process, while even economists with "outside" positions are *heard* within the councils of decision. Americans have come to the conclusion that, given their concern with economic security, they can ill-afford not to be Keynesians. Government's intervention in the economy is deemed prudent and productive, since most politically-significant sectors of society benefit from a stable pattern of ordered growth.

Galbraith's *The New Industrial State* recognizes the economist's essential status within the new order, as a planner and regulator of the economic stability and growth so central to the new system he sets forth.[11] Government and its economists must correctly intervene to guarantee an adequate and expanding demand for products. The economist, presumably integrated within the "technostructure," becomes an essential segment of the new State.

Americans have been preoccupied with their economic "health" since the collapse of 1929–1932. Economists have fulfilled the role of life-perpetuating physician to the economic structure. Even businessmen, long suspicious of government intervention in economic affairs, support the economists' and government's activities. The need for guaranteed demand and the self-interest of profit are apparently stronger forces than the need for ideological satisfaction received from residing in an Adam Smithian laissez-faire world.

For all these reasons, economists have been drawn into the policy-maker's circles. But they have not roamed freely with unlimited influence. Rather, economists have been influential insofar as their advice is needed, as they anticipate issues, or as they intentionally pursue a policy preference.[12]

THE COMMUNAL SECURITY ARENA

The communal security arena encompasses both foreign and domestic issues. But significant groups' perceptions of the domestic impact of even foreign-oriented communal security policies are very important. Weapons, deterrence and defense, health, and weather policies involve identical domestic impacts.[13] Policy-makers and citizens expect that the government will increase their common security, protecting them from disease, external aggression, or the caprice of the weather.

Government is expected to distribute security and continued well-being, particularly when these values are significantly threatened by outside or "natural" forces. The benefits distributed by government are usually perceived as communal, difficult to disaggregate to specific benefactors, and in the general or "public" interest. They are collective, insofar as all citizens benefit from a nuclear deterrent force, weather predictions, and mass polio vaccination programs.

The arena is quite stable and should remain so as long as the "cold war," disease, and the "elements" pose threats to the nation's and individual's security. Government must act to protect security because (1) it would be politically disasterous not to do so and (2) the security in question is also the government's own sense of security. When it comes to nuclear war, disease, and tornadoes, policy-maker and citizen are almost equally vulnerable, and therefore insecure.

Governmental activities may range from the distribution of military security to citizens to the regulation of drugs in the interest of public health. Formally, it may "regulate" (Food and Drug Administration), "distribute" (weapons system contracts), or be an "entrepreneur" (weather forecasting, modification). But regardless, the impacts are communal and increase security in varying degrees. Fiscal and monetary policy can also be seen in this light, as increasing common security against economic depression.

Domestically, communal security policies are nonzero sum, as most distributive situations are perceived. Still, though all members of the society benefit nearly equally, some sectors receive additional side benefits. This is particularly true in weapons and defense policies when the so-called "military-industrial complex" receives a windfall in weapons contracts, research grants, and jobs. In a similar sense, the medical sciences benefit from our concern with health, and agricultural interests benefit from government's weather activities.

Policies in the communal security arena are shaped on high levels of the Executive Branch, including the Presidency. This is particularly true for weapons and defense policy making. These benefits are common to all citizens, including policy-makers, and are highly salient and valued. Governments must protect physical security of life and limb before all else.

The government and its Executive Branch need scientists in shaping communal security policies, if only because the potential costs are higher. Mistakes are costlier in terms

The Communal Security Arena 181

of political support and national survival. Furthermore, the government is distributing security to itself (protecting itself) as well as its supporters.

But there are additional factors that supplement "need" and have given scientists *moderately high* influence within the Executive Branch. The relevant sciences are generally more technical, "scientific," and perceived as more "competent" by policy-makers. This reflects the subject matter these scientists study, since *things* are more manipulable than *people* (cf. weapons systems, disease, and even weather). But these sciences—such as chemistry, biology and physics— have been more bounteously endowed with research funds, a windfall not unrelated to the nation's need for their product and the security it would hopefully afford.

It is interesting to note that in health policy, precisely where scientists must deal with people, constraints do appear. Men want their health protected but will not readily change their life styles to achieve that end. Similarly, for defense policy and the Vietnam experience, men desired the security they thought it would bring, but for many it required a change in life style. This may explain some of Americans' oft-present desire to use the antiseptic, efficient atomic weapon to subdue quickly a small foe (Hanoi, Pyongyang) without costs of Americans' lives. So, too, do Americans see value in air power and firepower as a substitute for the foot soldier. Thus they achieve "more bang for the buck" and avoid changing life styles.

No vested interests hostile to scientists' contributions are predominant in communal security matters. "Security," as a value, is too widely shared, and its general support is widely recognized. Indeed, many special interests, such as the weapons industry, both benefit from and employ the scientists in question. Scientists themselves even have a vested interest. One wonders how much support for education and research stems from Sputnik and the threat harbored in the "cold war"?

Minimal division or indecision exists within the Executive Branch. Leadership support for scientists is prevalent. Scientific considerations shape many aspects of a policy. In the realms of medical science, chemistry, and physics, the sciences become much more difficult to comprehend. Secrecy insulates many communal security policies from public and even legislative involvement. Finally, a sense of urgency accompanies policy making in these areas, primarily because the stakes are so high.

Scientists, under all these conditions, have come to perform policy-making functions inside policy-making circles. They make recommendations and often prescribe and execute courses of action. Heavy reliance on contracting research with nongovernmental groups, such as universities and "think-tanks," becomes a pattern. Scientists have also been active in the media and in shaping public opinion. Such a role is particularly appropriate for "dissenters" and "outsiders" who oppose current or proposed policies.

Still, grouping health and weather with weapons and defense policies produces some strange bedfellows. The discussion so far has emphasized their similarities as "communal security" issues. However, they do have differences, which analysis of the individual cases will reveal.

Scientists in a Communal Security Arena: Weather Policy

The field of weather policy-making has typified the stable, relatively unchanging character of an old line, scientifically based, service-oriented government agency. From 1945 to 1968 scientists have enjoyed a moderately high level of influence over weather policy and policy making.[14] This continues despite increasing capabilities and programs for weather modification, an activity that increases conflict within the Weather Bureau and its citizen clientele.

Weather policy involves the scientific study, prediction, and even modification of natural atmospherically based events. It does not encompass various other environmental subjects, such as earthquake research and pollution controls, though they are housed within the Environmental Science Services Administration (ESSA).

Weathermen, or meteorologists, in predicting or modifying the weather, are likely to distribute or redistribute such values as security and wealth. They may distribute a sense of security to all citizens fearing surprise and harm by the weather, or they may serve special interests, such as agriculture or outdoor recreational enterprises. They may even redistribute rain from one area to another, selectively at given times, effectively bestowing wealth and/or misery on one geographical area and denying them to another area. This latter activity, increasingly a possibility, is likely to generate significant conflict, perhaps enough to convince large groups that weather policy is "redistributive" or "regulative" rather than "distributive" of communal security.

Competence and expertise are qualifications for policy-making positions in weather policy formation. Scientists, knowing full well that only they can adequately operate such a technical agency, have come to occupy most formal policy-making positions and functional stages of policy making. They have even developed a significant vested interest in the weather prediction and modification field, which coexists and benefits all groups that receive their services. That clientele would range from picnic-bound citizens to farmers harvesting crops to commercial air carriers.

Thomas Jefferson and Ben Franklin were both weathermen, and weathermen have a long association fulfilling a governmental function. They originally found relevance in the Army's Signal Corps (weather prediction for military maneuvers), later became more useful within the Department of Agriculture (farming), and finally moved into the Department of Commerce because their activities were increasingly relevant to business and commercial aviation. These strong clientele pressures that have supported the Weather Bureau through its history should be apparent, with the Farm Bureau and Air Transport Association currently being major interests.

Weather policy may be shaped by scientists, but their performance of the services they are expected to render is a politically sensitive matter. Failure to warn and specifically locate points of impending tornado impact has led to loud public criticism in recent years. Citizens want accurate weather predictions, but support from the citizens or clientele depends as much on the Bureau's failure to be right as it does on its day-in-day-out accuracy. Successes count for less than they might.

The political sensitivity of weather forecasting may be small compared to the political sensivity of weather modification. Initial experience with modification activities (rain making, snow making, hurricane diversion) indicates that the results, as might be expected, tend to please some and anger others. Forecasting increases security and wealth when right and distributes neither value when wrong. Modification regulates the weather to benefit some groups and to displease others. Regulative activities are more politically sensitive, particularly when the general public and not a single industry may be affected adversely by the regulation. And when the Bureau's activities are seen as "regulative" rather than "distributive" of communal security, scientists' influence would be adversely affected.

Government, through the Navy, Department of Agriculture, Bureau of Reclamation, and the Weather Bureau in ESSA, has moved only reluctantly in weather modification programs. The scientific base for modification remains uncertain and in its infancy. Agencies lack funds for weather modification activities and have little previous experience in such task-oriented programs. Indeed, some would prefer to transfer modification activities to mission-oriented agencies, leaving the Weather Bureau with research and forecasting responsibilities. Furthermore, the Bureau of Reclamation, emphasizing rain-making for arid western lands, disagrees with ESSA, which has other modification interests and bases of support.

Though a sizable private investment already exists in

weather modification operations (primarily rain-making), rationale for a primary role for government may exist. The area is so fraught with legal controversy (damages and blame related to modification activities) and concern for "communal security" that government may increasingly become involved in licensing or regulating the operations of private modification operations; resolving disputes over damages; or deciding which of its own agencies will have particular functions and responsibilities.

Weather scientists' activities have proved quite useful, though not central, to the pocketbooks and lives of most Americans. The general public has learned to accept the weather as "inevitable," except when someone tampers with it. Only the farmer, drought areas, industry, and recreational enterprises have mounted measurable pressure for weather forecasting and modification. As a result, scientist weathermen have operated in a developed policy-making climate, stabilized by long tenure as a government agency performing a traditional governmental function. Scientific components and considerations weigh heavily in weather policy-making, and under these conditions the "layman" or citizen is likely to defer to scientists, given his general approval of the Bureau's performance and impact of its actitvities.

Scientists in a Communal Security Arena: Health Policy

Scientists from the medical, chemical, and life sciences have status within the inner circles of the health policy-making. They have a moderately high level of influence in shaping health policy, thus affecting the personal and collective physical or mental well-being of American citizens.

These health scientists therefore directly and indirectly *distribute* well-being and security from disease and ill-health. But, in assuring this common security to citizens, health officials must often *regulate* the behavior of the food and drug industries and the behavior of the medical profession. This pattern of policy making persisted from 1945 through 1968, evidencing the stability of a function commonly ascribed to government.[15]

Health and medical scientists possess a high level of competence, monopoly on expertise, well-developed specialization and a heavy background in basic research. But they have moved beyond this scientific base into policy-making and administrative positions, accumulating a wide range of access points to the policy process and performing a variety of policy-making functions. In so doing, they have developed their own sizable vested interest in health policy and policy making.

Scientific considerations and scientists may shape health policy and dominate day-to-day health policy administration and research, but the services to which they contribute have significant political value. This is true for the communal security arena generally, because the affected values (communal security, life) are so important that no government could survive were it to fail in protecting or promoting them. Thus, if medical scientists perform their function and avoid health-policy mistakes that jeopardize security and life, the political interests of the government are served. And scientists are needed precisely because these values are so salient and common.

However, the health scientists do not have a blank check on policy making. Several severe constraints surround their activities and occasionally limit their influence, despite the general support health policy and the scientists have from political executives and the citizenry. These constraints include the following:

1. *Life Styles.* Though citizens are all for "health," they have implicitly imposed strict limits on possible measures intended to promote "health." Consider the cigarette smoking controversy, labeling of cigarette packages, and cigarette advertising. Here not only did scientists run head-on into a hostile vested interest in the cigarette industry and its

legislative spokesmen but also saw their general public support wither.

Indeed, the scientists did lack full research on smoking and health, but other factors have been equally relevant. The cigarette-smoking public, as in the problem of pollution, did not want to be deflected from ingrained, socially rewarded habits it had cultivated for so long. Smoking had become a way of life, its harm fully-rationalized, and many smokers were unable to see themselves in the statistics (cf. the value of wearing seat belts, pollution as a health hazard). Many smokers might well say, "You can protect my health, and I encourage you to do so with unlimited funds, but don't expect me to alter my behavior or alter it for me." Doctors must hear that position frequently from patients. Thus "health" may be a powerful symbol in politics that does not always translate to governmental action on behalf of actual health, particularly when the action would *regulate* citizens' behavior and force them to change their life style. Then the powerful symbol becomes a political stumbling block.

Citizens will not complain if special interests, like the drug industry, are *regulated* and denied the value of wealth from profits. Government can stop production and sale of Thalidomide because Thalidomide is not an integral element of a life style, but no health authority would readily make the citizen wear glasses, avoid fatty foods, or get a prescription for aspirin.

Again regulation of special interests is supported if it does not indirectly affect the life style of the general citizenry, but regulation of citizens' ingrained patterns of behavior can be politically disastrous.[16] Thus labeling of cigarette packages or bans on cigarette advertising might be accepted, because they are not seen to affect the behavior of existing smokers. But authorities would not ban cigarettes, or bring back Prohibition, without serious political risk.[17]

2. *Private Enterprise.* The cigarette and tobacco industries have not been the only corporate sectors affected by health policy. The politically powerful drug industry likewise

188 High Levels of Scientist Influence

has a special interest often threatened with adverse regulation. Drug manufacturers can tolerate significant government regulation of drugs as long as the testing and intervention only removes harmful items from the market. This is something the firms themselves attempt to do, perhaps without prodding from the Food and Drug Administration (FDA). The FDA and Federal Trade Commission may even lightly slap misleading advertising. Manufacturers then need only slightly alter their packaging and claims.

Some recent findings during 1970 do, however, indicate that the Food and Drug Administration has often modified or eliminated "adverse" scientific findings regarding the safety of food additives and pesticides. These irregularities apparently occurred within lower administrative levels of the FDA, probably in deference to industry interests or a desire not to precipitously rock the boat. Furthermore, a task force organized by Ralph Nader charged in 1970 that the FDA often distorted, altered, or ignored scientific findings regarding drugs. The task force felt that the FDA unduly deferred to "self-regulation" by the drug industry itself and was unduly dependent on the food and drug industry's own scientific research and scientists. Similar difficulties surrounded the "smoking" debate, when the Tobacco Institute had long held a monopoly on the government's ear for scientific research.

Finally, when health scientists and political figures threaten to tamper with drug prices (and profits) and the profit-laden patent medicine market, massive resistance and political pressure will come from the industry. Still, as long as public support can be generated behind such "regulation," the health scientists can exert some influence over policy. Special interests can be regulated much more easily than general interests.

3. *Congressional Appropriations.* Congressmen, who had long been willing grantors of funds and have blessed the National Institutes of Health (NIH) with an "embarrassment of riches," [18] late in the 1945–1968 period began to limit NIH funds. During periods of fiscal contraction, war pres-

sures, and budget cutting, one could anticipate what in fact occurred—an attack on the Institutes' budget practices, granting procedures, auditing practices, "unproductive" research, and favoritism.[19]

But the legislative assault, it is significant to note, was not on "health" as an objective of policy but rather on administrative procedures. For similar reasons, Richard Nixon in his 1968 Presidential campaign attacked only the administrative aspects of the popular Head Start program. "Health" without doubt remains a powerful symbol and policy objective.

4. *Medical Profession.* The interests and objectives of health scientists making health policy and of practicing physicians often differ, though on the Medicare issue they were not so far apart as the physicians were from the politicians favoring the proposal. Generally, the medical and health scientists confine their policy concerns to the scientific subjects of drug regulation, disease prevention, and regulation of health hazards. Medicare was more a social and political issue, and policy leadership came from legislators and the White House. But the pro-Medicare scientists and political leaders successfully defined the issue as the promotion of "health"—the powerful political symbol—and finally overcame the doctors' traditional definition of the issue as "socialism" or government intrusion on the doctor-patient relationship.

Government would distribute "health" care, even if it meant eventual regulation of doctors' fees. The American Medical Association, because its legislative campaign was an "all-or-nothing" effort, even failed to extract administrative responsibility for Medicare, a means by which the doctors could have maintained a *self-regulative* character in some remaining aspects of policy making.

Scientists in a Communal Security Arena: Defense and Deterrence Policy

Scientists from the fields of physics, economics, and systems analysis had by 1961 penetrated the inner circles of defense and deterrence policy making. Scientists from most scientific fields had been active within the media and public opinion from 1945 through 1960, but only a few could have been classified as internal policy-makers, administrators, or formal advisers. From 1960 through 1968, formal policy-making or administrative status partly eluded the scientists, and their main impact had to develop from "inside" advisory positions. Political and military leaders increasingly deferred to their judgment and analytical skills, often having to trust sometimes incomprehensible recommendations. This was the basis for scientists' moderately high influence during the period 1960–1968. The shift in 1960 will become apparent as the discussion proceeds.[20]

Defense and deterrence policy deals with a nation's general military posture and strategy, emphasizing military rather than disarmament or arms-control means to ends. Political, military, and scientific leaders responsible for defense and deterrence policy are presumably guided by the common preoccupation of government officials and citizens with their communal security and well-being.

The demands for government activities that maintain this sense of security come from all segments of the nation and, coupled with special interests' side benefits, present a mandate to government. A government cannot afford to err or fail in meeting these intense and common demands, and so it must use scientists whenever they might assure more security. Therein scientists contribute to the "distribution" of wealth to special interests, such as weapons manufacturers, defense research institutions, and military interests that benefit from the "warfare state."

Again, because fulfillment of demands for communal security is so important to the Executive Branch and Congress, scientists become relevant. Their influence has been moderately high, not because many specific issues turn on scientific considerations but because these men of physics, economics, and systems analysis have led defense policymakers to view the world differently, engage in comprehensive planning, and systematically account for values and factors relevant to the "communal security" goal. These contributions have profoundly affected defense and deterrence policy, even though in many respects the policy process remains a field of bargaining rather than reasoning.

Still, since 1960 the scientists have rendered defense and deterrence policy making less "politics"-ridden and more rational and comprehensive, arguing that their contribution has better enabled government to provide for the "common defense." "Politics" has been partially curbed to serve a larger political objective—increasing domestic security.

Scientists' participation and influence over defense and deterrence policy prior to the Kennedy Administration (1960) was not inconsequential. Because these experiences from 1945 through 1960 throw light on the post-1960 period, they might be briefly summarized as background material.

1944–1946. The failure of the Franck Committee to deter the use of the atomic bomb on Japan or adequately "educate" the public and decision-makers to the "foresight" the Committee possessed about the nature of international relations in an atomic world marks the beginning of the scientists' postwar history. The Franck Committee "doves" failed because the political leadership, public, and indeed the scientists on the Interim Committee's advisory panel (Arthur Compton, Ernest Lawrence, J. Robert Oppenheimer, Enrico Fermi) feared the loss of too many American lives in an invasion of Japan, thought a demonstration of the bomb might be an embarrassing "dud," or felt a demonstration would not have a convincing impact. Scientists might have succeeded in influencing policy had they been united, but they were not united; and the Compton–Lawrence–Fermi–Oppenheimer group strengthened the communal security-oriented predispositions of the political leadership in Washington.[21]

1946–1960. While the development of the hydrogen bomb was spurred by forces favoring weapons progress and superiority, many scientists who had opposed development turned to other strategies and efforts to influence defense policy. Oppenheimer became involved in Project Charles, later the Lincoln Air Defense Laboratory at Bedford, Massachusetts, working on continental aid defense strategy. Project Vista, led by Cal Tech's Lee Dubridge, emerged to urge development of a tactical nuclear capability and nonnuclear option for defending Europe.

An emerging breed of "strategic scientists" pushed for limited war capabilities, doing so in an environment dominated largely by physicists, chemists, and scientists with expertise in weapons development or weapons effects rather than by "scientists" specializing in strategy. The physicists, chemists, and weapons scientists had filled a void in the postwar period but operated essentially outside their scientific fields in dominating the strategy debates from 1945 through 1960.

Public debates during the 1950s over defense policy involved these physical scientists who acted not so much as defense-policy experts but as scientists who, like anyone else, held value preferences and strategic views and felt compelled to speak out. Many held opposing views on the Soviet Union's intentions, the need for shelters and civil defense, America's larger foreign policy purposes, the need for balance and variety in American military capability, the level of deterrence force necessary for security, the propriety of massive retaliation as a deterrence policy, and the importance of arms control measures.[22]

Policy from 1945 through 1960 did not reflect the "dove" scientists' positions, as groups of scientists failed to deter use of the atomic bomb, development of the hydrogen bomb, and Secretary of State John Foster Dulles' reliance on massive retaliation. Scientists developing and advocating weapons systems were generally influential, but only because the political leadership itself wanted all the hardware it could develop.

Still, the great debates on strategy, spurred mostly by physical scientists, served as catalysts for new strategic programs and criteria for defense policy that would flower after 1960. Scientists acted mainly within the media and public sector, largely because most were excluded from the inner policy-making circles. And they were abundantly successful in creating a climate in which strategic arguments were acceptable and "healthy." Debate became more respectable within and without the councils of government. This laid a foundation for the critical analysis of defense and deterrence policy during the 1960s.

The physical scientists during 1945–1960 had contributed to new definitions and perspectives for defense and deterrence problems. They posited new solutions to new problems, identified emerging dilemmas, urged changed goals, led public opinion, and appraised past actions. However, their actions took place more indirectly through the media than directly within the inner sancti of government. There was very little *science* behind their value-laden positions, but because they acted vis-à-vis a public that saw them as scientists, their advice was cloaked in an aura of scientific validity. Yet political leaders, building on the political wisdom of providing maximum communal military security and superiority, clearly shaped existing policy.

Some change occurred in this process after 1960. Defense and deterrence policy came within the purview of a wholly new group of scientists which Bernie Brodie calls "scientific strategists."[23] Their backgrounds ranged from physics (Herman Kahn) to economics (Thomas Schelling and Charles J. Hitch), and they deliberated with logic and systems rather than with military experience and common sense. Their use of systems analysis incorporated a scientific tool into defense planning, but more importantly they injected scientific attitudes, such as skepticism, logic, and need for empirical tests into policy making. Ralph Lapp, observing a "pronounced military orientation" in these and other defense scientists, explains the behavior of many by saying they grew up in the defense establishment since World War II and failed to experience the "open world of free science by submerging themselves in secret activities." [24]

Full-time concern with a "scientific strategy" thus became an applied social science and replaced the makeshift arrangement during the 1950s when physical scientists fulfilled the function of strategic analysis on a part-time basis. Still, however, some scientists waged their battle in the media during the 1960s (cf. Linus Pauling, Erich Fromm), and political and military leaders rather than scientists continued to occupy formal policy-making positions.

The new "strategy scientists" perform as advisors with status within inner policy-making circles and produce needed critical analysis and comprehensive programming. Their contributions brought on policy innovations (counterforce doctrines, flexible response strategies, guerilla capabilities, arms control measures) and instrumental decision-making innovations (systems analysis, comprehensive planning, cost-effectiveness rather than the old budget ceiling and single-item approach to decision making).

Political leaders like President Kennedy and Secretary of Defense McNamara supported the participation of the "strategic scientists" while military men, professional diplomats, political appointees, and traditional scientists moved toward the periphery of the inner advisory circle. Merging civilian and military criteria for decisions, these new men gradually developed their own vested interest in defense and deterrence policy making.[25]

While serving their own self-interest (prestige, jobs, a sense of shaping *military* policy), these academicians or scientists increasingly served the interests of private enterprise, "nonprofit" research organizations, and the military. They found usefulness for wider ranges of weapons systems and generated a need for vast quantities of defense policy research.

But research and the "strategy sciences" were not unnecessary. Their usefulness stemmed from the fact that weapons and strategy decisions had been pushed into the prewar stage with the advent of nuclear weapons and missiles. Decision times were cut to the minimum, rendering planning and contingency targeting indispensible. Military experience might be valuable for fighting a war or mobilizing after the outbreak of a war. But when men must plan for a type of war that never has been fought, experience becomes a deadweight. Increased awareness of the potential enemy's own freedom of action has also made game theory, war games, and simulations more useful. Scientists have responded with research in these fields. However, the main point is the advantage of logic and comprehensive thinking when decisions are pushed into the hypothetical prewar period during which all weapon selection, design, and deployment must occur.[26] Post-attack mobilization would be foolhardy,[27] particularly in the nuclear era when "the value of combat experience seemed drastically depreciated." [28]

The economic and systems sciences were most useful to the "strategic scientists," though the social scientists also became more involved during McNamara's tenure as Defense Secretary. The economist's conceptual framework, inherited from Adam Smith, was well adapted to strategy decisions because it placed a value on all costs and benefits and permitted choices normally produced by the market mechanism. This usefulness for policy making is probably due more to chance than to wisdom, eclecticism, and foresight from economics' founding fathers. Specific examples of derived techniques have included:

1. Cost effectiveness and cost accounting; Planning-Programming-Budgeting System (PPBS) developed by Charles J. Hitch and Alain Enthoven under Secretary McNamara; program budgeting.[29]
2. Simulations; war games; game theory; conflict studies;

scenarios (cf. Herman Kahn's *On Thermonuclear War* and Thomas Schelling's *Strategy of Conflict*).
3. Rand Corporation's strategic bases study during the 1950s under Albert Wohlstetter, a study effectively incorporated into Air Force policy.[30]
4. Computer systems such as SAGE (air-missile defense).
5. Research and analysis of the effects of thermonuclear weapons.
6. Systems analysis and the newer "policy sciences" as more comprehensive tools than the operations research conducted in World War II, which was tied to particular low-level tactical and supply problems (new concern with strategy mixes and optimizing comprehensive multivariable planning).

But the "strategy sciences" wrought more subtle changes in defense policy-making, perhaps by infusing an attitude and spirit into the process. Systems thinking forces consideration and weighing of all factors in a decision; game theory suggests consideration of an opponent with bargaining power and self-will; and simulations and scenarios encourage thinking about the future in constructing any strategic equation.[31] These developments were not fully present in the 1945–1960 period.

However, regular and "strategic" scientists have never had a clear field for exercising influence over defense and deterrence policy. Four basic factors limited their influence in varying degrees:

1. *Scientists' Conflict.* No doubt scientists' influence suffers from the high levels of political and value-based conflict scientists have with one another. Indeed, while the "strategic scientists" have less conflict than their nonspecializing counterparts in the physical sciences, the infancy of their "science" still harbors value-based and scientifically based disagreement. There are even differences over the merits of systems analysis and cost effectiveness. Political leaders cannot help but depreciate the scientists' advice

whenever it is bogged down in scientists' own internal disagreements.

2. *Congress.* Congressmen concerned with military. expenditures and defense matters share the general hostility of military men to the new innovations in decision making. As such, this constitutes some limitation, even though the new techniques flourished despite these legislative reservations.

3. *Political Tradition.* Defense and deterrence policy remains a political and civilian responsibility exercised by political leaders. Indeed, successful fulfillment of demands for communal security is politically beneficial to a leader, and so political leaders have maintained their prerogatives, deferring only when necessary to military men and scientists. It has become increasingly "necessary," however, for leaders to defer, since the outbreak of war or threats of nuclear destruction are sufficiently serious threats to security that any contribution to meeting them, whether from scientist or general, is welcomed.

4. *The Defense Policy Process.* Defense and deterrence policy develop from a process laden with bargaining and compromise among diverse and competitive forces. Pressures increase the likelihood that change will be only incremental and consensus-bound. The process retains the legislative characteristic of equals bargaining among equals. Tendencies persist to paper over conflict, delay decisions, express policy in vague generalities, find lowest common denominators, roll logs, make decisions by committee, avoid coordination, and avoid anticipatory policies.[32] These conditions are detrimental to the synoptic contributions of scientists.

Samuel Huntington describes the usefulness of scientists this way:

> The desirability of innovation does not become equally visible to all groups simultaneously. A group can take the lead in breaking the previously existing consensus, or in "creating an issue," only if it is both sensitive to the need and also free to propose and push programs to meet the need. . . .

198 High Levels of Scientist Influence

> The initiative in program innovation, as in weapons or organization innovation, often rests with the civilians or military who are "unconventional" or "outsiders." The smaller the stake which a group has in existing programs, the more likely it is to push new programs.[33]

However, innovators must compete with other innovators and noninnovators, and scientists must compete with other groups. Clearly:

> For favorable action on an advisory recommendation to occur, the adviser must normally persuade large numbers of people throughout the decision-making organization and foster something of a consensus among the interests affected by the recommendation. In short, "closed politics" seems to resemble "open politics" in a number of important respects.[34]

The need to compete with political and military leaders who have their own objectives therefore would seem to be a limitation on scientists' influence.

Scientists in a Communal Security Arena: Weapons Policy

Weapons policy deals with the development, production, and deployment of specific offensive and defensive military hardware. Scientists have always developed weapons for men; but over the period 1958–1968 scientists' participation in weapons decisions became much more involved and perhaps sinister compared with early gunsmiths or Archimedes and da Vinci developing engines of war, bridges, mortars, and catapults. A distinct breed of scientists building their careers around military hardware has emerged. And with the Kennedy Administration and Secretary of Defense McNamara, many other types of scientists were brought into weapons decision making.[35]

Scientists active in weapons policy decision making include individuals from the fields of applied science and engineering, physics, economics, and systems analysis. Their activities are shrouded behind the invisibility generally accompanying national security decisions. The values affected by these scientists and weapons policy relate primarily to "communal security" and the nation's sense of well-being (freedom from external aggression). The government and citizens alike share in this distribution of security in the form of weapons and military superiority.

But there are side benefits that assist the efforts of scientists favoring weapons development. Weapons contractors receive profits and wealth. Political leaders derive a sense of personal and electoral security. Working men on defense projects obtain increased income from jobs. Scientists themselves derive prestige, research funds, and perhaps even a sense of discovering scientific "truths."

No large group in society is directly harmed by weapons expenditures, even though these expenditures may siphon funds from ghetto programs and thus indirectly affect the poor or other groups making demands on the government's resources. The scientists' activities enjoy a broad base of support, including the support of political executives in the Pentagon and military men who must procure weapons for their "trade."

Scientists have become involved in the inner circles of weapons policy making and have several points of access for influence. Scientific considerations shape the development, production, and deployment of a weapons system. Considerations *favoring* the weapons system are given more weight and attention, however, than considerations suggesting it should not be procured. Only scientists' argument that the system will not work when in use seems to have a negative impact on development, and even that may be disregarded by military and civilian leaders preoccupied with developing any remotely possible system.

This pressure to "buy everything" stems in part from the universal demand that the government achieve weapons superiority on every level in order to guarantee "communal security." More research and production benefit the scientists and manufacturers, too, and this makes them that much less likely to oppose development of systems. Therefore, conflict among scientists is often disregarded in a weapons decision. The view of pro-development scientists is tentatively adopted, and the process toward deployment proceeds.

Scientists involved in weapons policy making do not alone make authoritative decisions to develop and deploy weapons systems. They do determine feasibility, build the initial device, and supervise its production and deployment. Some, either economists or systems analysts, use systems or accounting techniques to choose among weapons systems or mixes. Overall, scientists' influence on weapons policy is moderately high.

SCIENTISTS AND DEVELOPMENT. The scientists' acts of discovery or determination of feasibilities may in themselves foreordain weapons decisions. The scientist who knows how to build an atomic bomb or antiballistic missile system has in effect already determined the final outcome, given the United States' concern with military security and weapons superiority. Choices from this point of discovery or innovation forward tend to be decisions of priority rather than "yes" or "no" choices. Innovations and ideas contain a built-in momentum, which propels them into reality and contemplation of their military use. Einstein's letter to Roosevelt proposing development of the atomic bomb, Sir Henry Tizard's insistence on radar, Edward Teller's work on the hydrogen bomb, John von Neumann's successful miniaturization of a hydrogen bomb small enough for an ICBM payload, and Werner von Braun's development of the Saturn rocket for moon exploration or military use illustrate the momentum process.

But these scientists' influence is not relegated to discovery per se. They also may actively urge political leaders to develop and deploy the idea. They may support the weapon's development or counter its critics with scientific arguments as it proceeds toward deployment. They may actively seek increased research funds to refine their design. But these scientists have had little influence over the *use* or *nonuse* of their weapons system, either because they are excluded from defense policy-making or because they abdicate responsibility for its use. Dr. Louis F. Fieser, leader of the Harvard team that developed napalm during World War II, typifies the abdicator's position in contending:

> That wasn't my business. That is for other people. I was working on a technical problem that was considered pressing. . . . It's not my business to deal with moral or political questions.[36]

Exclusion and abdication from responsibility for weapons use is illustrated in the case of the atomic bomb, itself an example of the "momentum" in weapons policy decisions:

CASE IN POINT: THE ATOMIC BOMB.[37] Leo Szilard and Eugene Wigner's initiative, backed by the Einstein plea to President Roosevelt, overcame existing inertia even within the military. When some feasibility glimmered in the darkness, the temptation to build (and, perhaps, use) the bomb was irresistible. The original group of Hungarians (Szilard, Teller, Wigner) had acted on highest motives founded in their personal experience with the Third Reich as refugees and their strong desire that the Allies develop such a bomb before the Germans.

Still, military men then felt that morale rather than weapons won wars and initially doubted the usefulness of such an enormously powerful weapon. But they acquiesced to the persistent and dedicated efforts of the scientists. These scientists' actions made deployment and use much more likely and rendered the series of events "irreversible." They cut through governmental red tape and overcame the alienation and lack of confidence that had permeated their prewar relations with political leaders.

The scientists were the main force behind the bomb's development.[38] They developed the device, and it became the "property" of political and military leaders making defense policy. Scientists were relatively excluded from this latter process. They were sharply divided over the military use of the weapon and the specific manner in which it would be used. One group urged a demonstration or nonuse. Another concluded it was wiser to explode the device on a Japanese target. Lewis Feuer has observed that this latter group of well-known scientists favoring military use held a "hardline" position:

> Among the various groups which had a voice in the decision to use the atomic bomb, the official leaders of science were the least humane. The military chiefs, oddly enough, showed the greatest human compassion.[39]

Thus, some scientists favored the bomb's use, some favored its use in a demonstration, some were excluded from decisions on its use, and some abdicated from such political and military decisions.

SCIENTISTS AND CHOICE. Scientists, notably economists and systems analysts turned defense planners, have had some impact on weapons decisions. They have attempted to choose among or rank different systems depending on their costs, usefulness, and the added security they might offer. These applications of systems analysis and "cost-benefit" or "cost-effectiveness" techniques led to cancellation of some weapons systems in the Kennedy Administration under Secretary McNamara. Poseidon, B-70, Skybolt, the 100-megaton bomb, Dynasoar, ABM, Minuteman III, and MOL developments were cancelled or indefinitely postponed.

The techniques also assisted choice among various weapons designs (cf. selection of a TFX/F-111 contractor). However, two factors depreciate the role of economists and systems analysts. First, many of the "cancelled" systems bounce back in new forms or are revived after dormant, continuing research periods (cf. ABM as the "Safeguard" system and reinvigorated R&D on the B-70, B-1, or AMSA under the Nixon Administration). Second, systems analysis has erred because it could not accurately anticipate the future (cf. cancellation of the armed helicopter in 1959, prior to its need in Vietnam-style conflict) or because it was used to mask or "justify" essentially political decisions. Stanley Jarolem and Daniel Kahn, writing in the *New Republic,* explain the latter failure:

> The gravity of the F-111 program is not only the failure to produce the aircraft specified, but also that serious questions are raised with reference to the use—or non-use—of systems analysis. It seems clear now that a rather subtle attempt was made to justify the original decision of design and contractor on the basis of sophisticated calculations. It is now equally clear that under the guise of using these sophisticated calculations some other method, yet unknown, was utilized to make the decision.[40]

Robert J. Art's case study of the TFX/F-111 decision suggests that officials' differing needs and objectives, coupled with McNamara's desire to control the military's role in weapons decision making, were basic factors shaping the TFX outcome.[41] But these are more political than scientific criteria, even though systems analysis did lead to McNamara's demands for "commonality" in design for multi-service use. To some extent, the new decision techniques have increased control over a weapons policy predisposed to "buy everything." Their role and impact has been to raise the question of "choice," a function the techniques are designed to serve.

Scientists have a moderately high level of influence over weapons policy not only because they can build weapons systems or assist choice among weapons systems but also because conditions inherent in the policy process facilitate that influence. These include the weapons complex, the incremental nature of weapons decisions, the complexity of weapons systems, and Americans' preoccupation with weapons as sources of security and substitutes for commitment of human lives.

1. THE WEAPONS COMPLEX. Scientists' contributions are needed and supported by a permissive Congressional attitude

toward defense expenditures, contractor pressure for development and ample supplies of contracts to maintain profits, labor's need for jobs in defense-impacted areas, and an Executive Branch concerned with the communal distribution of security and wealth to citizens and industries.

Scientists themselves are members and beneficiaries of this amalgam of interests. Scientists with access to classified documents and policy-making circles have more interest and basis for influence than scientists without access. "Insiders" can criticize decisions, while "outsider" scientists must criticize after decisions are made, a position comparable to resisting a coup d'etat after the fact. And "insiders" who are deeply involved and identified with the weapons system being analyzed are reluctant to criticize or develop antidevelopment or deployment arguments. Scientists not deeply involved with the weapons system, however, have no information on which to base criticism.

Scientists engaged in basic, nonapplications-oriented research are excluded from the policy process, an exclusion that Project Hindsight seemed to suggest.[42] Defense industries' scientists, on the other hand, generate scores of ideas and designs. Indeed, "defense firms in particular are prolific sources of evolutionary ideas . . . their supply of improvement ideas far exceeds the demand." [43] So close are these relationships between defense contractor and government that it is hard to tell where one begins and another ends, or just how a firm differs from a traditional government corporation or bureau. Private becomes public, and public becomes private.

These vested interests of labor, industry, military groups, local communities, politicians, administrators, and scientists forge a hospitable climate for scientists whose contributions sustain these vested interests' benefits. The "weapons culture" continues to derive wealth, and the entire American culture derives communal security. Jobs, prestige, wealth, security, and a preferred life style are sustained and made secure by the production and proliferation of weapons systems. Ralph

Lapp sees this as an *institutionalized* culture in which men view their participation as a "job" not involving questions of individual or collective responsibility for the results of their work.[44]

So strong is the "culture" that Congressmen can often override scientists' recommendations against development, as they did in the case of the nuclear submarine. Scientists attempting to prevent development or deployment of a weapons system are invariably excluded from policy making and can get formal access only through the President's Science Advisory Committee (cf. scientists' negative feelings on nuclear-powered aircraft and the B–70).

The close Senate vote in 1969 on the Safeguard ABM system's deployment indicates that some Congressmen are challenging the "weapons complex" and giving opposition scientists a new avenue of access and influence. But even in that debate, scientists' arguments that the system would not work as designed met proponents' faith that any problems would be overcome in refinements as the system was deployed. As a result deployment was authorized and begun.

2. INCREMENTALISM. Peck and Scherer see the weapons policy process as incremental insofar as:

> . . . literally hundreds of individuals participate directly in the typical weapon system program decision, and the result is a lengthy and complex decision-making chain. Almost inevitably the chain will include both advocates and skeptics, and an important dissent may seriously delay a program considered by others to be urgent.[45]

However, though dissent may *delay* a program, it far less often leads to cancellation of a program. Secretary McNamara did cancel the nuclear-powered aircraft program on the advice of the President's Science Advisory Committee (PSAC). But he was able to do so because science simply argued that existing aircraft technology would not enable lifting a reactor off the ground unless an efficient high temperature reactor was developed. A nuclear powered aircraft would be impossible for some time to come, though it is still no "dead duck." The B-70 bomber, opposed by PSAC and Science Adviser George Kistiakowsky, was cancelled.

Still, what often looks like a cancellation may only be hibernation and *delay*. The pressures for development still exist, and new technologies might anyday resurrect the projects. Short of outright "impossibility," weapons ideas have a "momentum of their own," which will ultimately shape and constrict a nation's defense and deterrence position. Weapons decisions are priority decisions, David Tarr persuasively argues, and fit nicely into the military arena characterized by logrolling, compromise, and a "buy everything" psychology.[46] Each demand for a weapons system is met, at some time or another, because there is no limit to communal security and because weapons systems are the "coin of the realm" in the back-slapping atmosphere of the weapons "culture."

Similar situations surround scientists' efforts in developing chemical and biological warfare capabilities; tactical computers, infra red gunsights and fragmentation bombs for guerilla conflicts; large megaton strategic bombs; nuclear and conventional tactical weapons; or scientific devices or techniques for logistics and management functions. Scientists contribute desperately needed skills, which, if David Tarr is correct, have great consequences for defense and foreign policy. Tarr notes that:

> More often than not weapons grow out of the impetus of what is technically possible—especially major or "strategic" weapons systems—and their military and political values tend to be defined after the fact.[47]

3. COMPLEXITY. Political executives and military leaders who must deal with the scientists' contributions must face the extreme complexity of the sciences and weapons systems these scientists create. Don K. Price, building on this assumption, foresees an ominous possibility and contends that the:

The Communal Security Arena

> ... danger is probably that the top political authorities of the government might find themselves in charge of a system which gave them too little flexibility, and too little range of choice in difficult diplomatic or strategic situations, because it had been based on decisions by engineers and military officers which the politicians did not fully understand.[48]

Scientists' influence then is augmented by the difficulty nonscientists have in comprehending their science, because their advice and judgment cannot be adequately tested or questioned.

4. THE NEED FOR SECURITY. Scientists could not design the weapons systems they design or successfully influence their deployment if Americans did not believe that "communal security" is produced by the process. The political climate in the United States, with a public that demands security and sees weapons superiority as a means to security, renders scientists' proposals irresistible. Weapons policy-making may be invisible and hidden from popular scrutiny and democratic control, but there is no real evidence that most Americans would want to control or change a process that has produced satisfactory levels of security in the past.

Scientists may not themselves make weapons choices (if choices are really made, or if any one person makes them), but they do in fact influence outcomes. They do so by posing possibilities to military men anxious to fulfill their function and citizens occupied with security. Weapons seem to be a means to security, and we cannot resist the temptation to become more secure. So strong is our concern that we are willing to "buy everything" in the weapons line, even if we are behaving uneconomically and escalating our costs. We must close every security loophole and preempt every contingency.

But we do not need weapons merely to fulfill a psychological need for "security." The reasons are far more complex. Indeed, foreign enemies may develop and deploy a given weapon first, when the United States has neither that weapon nor a defense against it. Also, perhaps we develop weapons for the same reasons men climb mountains: because the possibility is there. All of these explanations converge in the decision to develop the hydrogen bomb.

CASE IN POINT: THE HYDROGEN BOMB.[49] Though prior to 1958, the development decision heavily involved scientists on both sides. The Atomic Energy Commission's General Advisory Committee (scientists Lee Dubridge, I. I. Rabi, Robert Oppenheimer and Cyril Smith with Enrico Fermi dissenting) opposed development of the hydrogen device on political and scientific grounds. Many argued that it was not feasible, or at least not feasible through the methods then being used. However, Edward Teller, Luis Alvarez, and Ernest O. Lawrence favored development, and President Truman agreed to continue supporting their research. Truman's decision, though minimal and closing off the least number of future alternatives, was the only political choice.[50]

No President could have decided against continued development at a rapid pace, despite scientific dissent, because public and Congressional pressure for "security" would have erupted in politically harmful protest.

Scientists opposing development extracted some concessions in the National Security Council's general policy review that led to the NSC-68 document. But the decisions in NSC-68 resulted in new weapons developments, opportunities for scientists frustrated with the apparent reliance on strategic nuclear weapons. They became involved in Project Vista (tactical nuclear weapons), Projects Charles and Lincoln (continental air defense), and Project East River (civil defense and urban dispersal).[51]

Weapons achieve "communal security." Even though they are costly in income taxes, they are an easy technological path to security. Weapons policy, like health, social, pollution, and other policies, finds public support dwindling when policy requires individuals to commit their lives and time or change their styles of living. An antiballistic missile system (ABM/Safeguard) meets these popular objections. Civil defense, requiring active citizen participation, does not. Technological paths to security are psychologically more reassuring than diplomatic and personal paths. Air power is more antiseptic and less "dirty" than ground warfare, and it costs less American lives.

Americans would prefer to substitute hardware for "humanware" in any war, even if it means the use of nuclear weapons. Americans tire from the human costs of war, not the overall costs of equipment and weaponry. Scientists' contributions enable attainment of this impersonal, technological, and undemanding road to citizens' and governments' common security.

THE ENTREPRENEURIAL ARENA

The Executive Branch's entrepreneurial activities stem from broader concepts of the activities of government, the lack of private capital or sufficient profit in selected activities, or new demands thrown up by a technological society. Government, in any case, becomes the sole entrepreneur capable of commencing and sustaining some activities in the society.

The impact of entrepreneurial activities is distributive and often borders on being pork-barrel politics. Wealth, enlightenment, skill, well-being, or security may be distributed, either to special interests or the communal interest. But, in any case, groups do not see other groups' benefits as coming from their pocketbooks (redistributive impact).

The activities are allocative in character, even though there is a basic uncontested structural assumption behind government's entrepreneurship—that government rather than the private sector will develop these fields.[52] The structural assumption goes uncontested because (1) private enterprise may not have the capital to finance such vast activities, (2) profit margins or risk may be very high, (3) the "product" may not be marketable like soap because it is a community benefit like military defense, (4) the activities are in the government's own interest, or (5) no private groups, which might have been harmed by the intrusion, were involved in the activities before government entered the field. Therefore, the game is perceived by politically significant groups to be nonzero sum or distributive.

Entrepreneurship by an Executive Branch may take varied forms. Government may undertake the economic development of undeveloped regions (Tennessee Valley Authority). Government may actually produce a product from raw materials, either in its own laboratories or through contractual arrangements (Atomic Energy Commission, Army arsenal system prior to "contracting-out"). Government may contract for finished materials from private enterprise and then use these materials for its own entrepreneurial purposes (space exploration and uses of space). Finally, government may deem a field so important to its own needs and the community's continued security that it bankrolls individuals' and groups' efforts with vast monies, which would not be forthcoming from private sources (national science policy and scientific progress, ocean exploration and utilization).

Entrepreneurial policies are centered in agencies and have strong support on higher executive levels. Congress participates in a supportive manner, although it may occasionally affect the aggregate level of entrepreneurship in a field by imposing budget ceilings or chopping total budget requests. This legislative support stems from the distributive character of the activities and policies' contribution to the national interest (security, economic growth, etc.).

Most governmental "entrepreneurial" efforts were stimulated, and continue to derive support from, the "cold war" and demands for "communal security." This is clear in the fields of space policy and science policy, as brief characterizations will show. National prestige is also not unrelated to rising involvement of government as an entrepreneur.

Entrepreneurial activities, because they are new endeavors and because they are a response to the technological society, tend to involve scientists in most operational or policy-making positions. As a consequence, scientists will be moderately highly or highly influential. They will lend influential and complex advice, determine policy objectives, decide matters affecting their own research support, administer programs, make compelling recommendations, and engage in public debate.

Scientists' desire to exercise power and influence policy may be greater in these "entrepreneurial" fields than their desire to discover scientific truths. Still, their status obviously depends in the long run on their scientific productivity. Scientists may even be influential simply because they have interests affected by space and science policy making. These interests may be so substantial that they will manage their own pork-barrel distribution system (science policy).

The most active and influential scientists in entrepreneurial arenas are physical scientists and, to some extent, biological scientists. Social scientists are significantly less influential and involved.

Support comes from private enterprise, which benefits from scientists' needs (space exploration) or the scientific knowledge they produce (scientific progress, funding for scientific research). Scientists also enjoy support from political executives in the Executive Branch and from legislators, even though Congress may sometimes trim their aggregate level of support. But even with a smaller budget, scientists make policy affecting the distribution of that smaller total pie.

The scientists are influential in part because only they have the needed skills. Such skills are required and supported by government because government sees "entrepreneurial" activities in its own interest and because so many of the activities appear to protect or augment communal security. Indeed, massive federal support for education, scientific research, technological progress, and space activities can be attributed to concern with national security and "cold war" competition with the Soviet Union and China.

Scientists' participation in space and science policy making will illustrate and clarify their influence over the Executive Branch in the entrepreneurial arena.

Scientists in an Entrepreneurial Arena: Space Policy

The government, through its National Aeronautics and Space Administration, is an "entrepreneur" in outer space. To even begin activities in such a highly scientific enterprise, scientists are necessary as advisers and policy-makers.

But science and scientists benefit from government's activities. They receive research funds, facilities, jobs, prestige, skills, and enlightenment as a result of the largesse flowing from the space program. No small wonder that they consider space policy an important field for access and influence. Scientists' own self-interests, society's need to have skilled manpower to design and operate such a highly technical program, and the benefits accruing to government and private industry ensure that scientists have a moderately high level of influence over space policy-making within the Executive Branch.[53]

The space effort did not significantly begin until 1957 when the Soviets launched their Sputnik, and so this analysis deals with the period from 1957 through 1968. This fact in itself indicates that scientists' influence is not all their own making, but rather depends on Americans' desire to maintain world prestige and their present level of security.

But with an established space program running, other factors come into play. Industries benefit from NASA contracts. Political leaders' prestige rises with successful space exploits. Funds for education are increased. Scientists bask in untold riches flowing from a barrel of pork that, in large part, they are in charge of distributing to themselves.

Space policy relates to the development, exploration, and analysis of the earth's environment through the use of orbital and interplanetary technologies. It does not deal directly with military uses of space, though it is obvious that threatened military uses of space and rocketry spur a nation's "civilian" and "peaceful" efforts in space.

Despite scientists' moderately high influence within the Executive Branch and NASA, and their occupancy of advisory, policy-making, and administrative positions, they have not used much scientific research or science as tools aiding policy making. Scientific judgments may shape individual decisions on missions and hardware development, but rarely do the scientists use science to select among general objectives and allocation of funds. Instead, the field is dominated by bureaucratic, political, and personal motivations. This is one reason why scientists recently raised cries of alarm over post-Apollo funding. Jobs, research funds, and prestige took precedence over dispassionate, scientific analysis of the merits and demerits of post-Apollo activities.

Scientists' predominance in space policy making was first boosted by the Soviet Union's orbiting of Sputnik in 1957. Immediately America resembled a world of Jules Verne:

> The immediate results of Barbicane's proposition was to place upon the orders of the day all the astronomical facts relative to the Queen of Night. Everybody set to work to study assiduously. One would have thought that the moon had just appeared for the first time, and that no one had ever before caught a glimpse of her in the heavens. The papers revived all the old anecdotes in which the 'sun of the wolves' played a part; they recalled the influence which the ignorance of past ages ascribed to her; in short, all America was seized with seleno-mania, or had become moon-mad.[54]

Americans were challenged by the Soviets and taunted with the possibility of escaping the earth. Scientists were given wide latitude, assuming that their activities would restore lost prestige, assure American control of space, and incidentally overcome the threat of Soviet missiles to Americans' military security. Science could provide a *technological solution* to the loss of prestige, insecurity, and desire to escape into outer space.

214 High Levels of Scientist Influence

The latitude and largesse of funds would continue for a decade until inflation, the costs of Vietnam, urban needs, and popularity of the budgetary axe skimmed off some of the cream. In 1969, NASA appropriations were down 20 per cent from 1968, leaving only sufficient funds for a series of Apollo moon landings and forcing delay in basic research-oriented and planetary missions. Even the successful moon landings of Apollos 11 and 12 did not stimulate an upsurge of funds. A shift in priorities had perhaps begun.

Beyond the constraints of budget-cutting and competing domestic demands, there are other limitations on scientists' activities. Public opinion recently shifted from a "blank-check" policy and has expressed some doubts about the value of the space program. Opinion does, however, tend to recover and support space activities whenever a successful space "spectacular" occurs. It falls when failure strikes, as in the Apollo fire killing three crewmen on the ground. But, generally, the public has become less fascinated by space activities in the past decade.

Scientists' activities, however, are also limited by national political objectives. Space exploration is a "political" as well as "scientific" goal. Indeed, though some scientists felt that manned exploration of the moon was scientifically unjustified, most agreed that such a mission would generate more political support. The Saturn rocket, conceived and developed for possible military use, was moved into NASA's moon program in 1959 when it no longer seemed to have a military function. Space scientists' more prestige-oriented and military-oriented programs, most observers would agree, generate political support and allow less popular, purely scientific programs to piggy-back along.

So strong were the demands for regaining prestige and assuring scientific and military superiority that President Eisenhower's deliberate attempt to play down the Sputnik threat and America's response met with failure. President Kennedy, faced with a highly technical debate over the best means to reach the moon, relied on his "political" senses to make a decision.

NASA Director James Webb and Werner von Braun had favored a single Saturn shot to the moon, orbiting the moon, descending and rendezvousing for return to earth. This course was riskier but could be accomplished earlier. Kennedy, sensing the political value of beating the Russians to the moon, chose that course. He rejected the views of his Science Adviser Jerome Weisner and the Pentagon's Harold Brown, who had argued for a more scientific course with limited military value. Under their plan, astronauts would assemble a powerful rocket in earth orbit, descend to the moon and return without orbiting the moon. This course would have been safer and enabled development of techniques for satellite inspection but might have allowed the Soviet Union to reach the moon first.[55] Then, too, so important was prestige and scientific progress that criticism by scientists of the moon objective was minimal. No one seriously considered not going to the moon, later or at all.

In some respects the United States has no policy for space affairs.[56] What appears to be policy may be a series of spur-of-the-moment decisions. Meg Greenfield has reported that the moon program was worked out one weekend by NASA Director Webb and Secretary of Defense McNamara as a response to the impending economic recession in 1960–1961. Congress would not support spending on domestic social programs, but political leaders felt the legislators might buy the moon as an economic stimulus for the economy.[57] However, the fact that no clear space policy exists leaves the field open for scientists' to dip into an ad hoc "pork barrel." An absence of clear policy itself may produce outcomes just as laissez-faire economics leads to certain outcomes. Some scientists inherit the vacuum.

Although the space "establishment" remains relatively united, both scientific and political issues have divided scientists concerned with space activities. Nonspace scientists and some biologists, frustrated with a physics-oriented space program, constitute a vocal group of outsiders who cannot critically attack policy because they lack full-time experience and current knowledge of space science. Full-time "insiders" possess a high level of competence, significant vested interest in continuation or expansion of the program, and a monopoly on knowledge on this fast-developing and complex scientific field. Criticism from "outsiders" is therefore often irrelevant or wrong, and "insiders" have no interest in publicly criticizing their own efforts. Thus "insiders" make policy.

Space scientists' influence rests on other factors. They have in a sense created their own demand by raising the possibility of space exploration, though here science-fiction writers have perhaps done more than anyone else in stimulating men's imaginations and ambitions. Scientists have stressed the spin-off and domestic usefulness (medicine, computers, electronics, intercontinental television, navigation, missile technology) to further justify their activities. Finally, they have argued that space exploration is another of man's steps for scientific progress and conquering the unknown. Indeed, man often feels that knowledge is intrinsically valuable and that mountains are meant to be climbed.

Scientists in an Entrepreneurial Arena: Science Policy

Science policy means "policy for science" and involves such activities as research funding, support of university science, and efforts to increase the United States' scientific capabilities. Government, deeming these objectives so important, has in the postwar period, and particularly since the launch of Sputnik in 1957, poured vast funds into scientific research.

Support has been so extensive that Washington has become the major scientific "entrepreneur", supporting research that no other group could or would support. In so doing, government has built up the nation's communal store of enlightenment, scientific skill, scientific manpower, and sense of security in an educational and scientific race with foreign powers. Indeed, support of scientific research can be attributed as much to the "cold war" as to scientists' own demands for support.[58]

The Executive Branch, though acting as an entrepreneur, is in effect "distributing" values like skill, security, and enlightenment. At times, this "distribution" resembles a pork barrel for scientists, and scientists are given the privilege of dividing the pork among themselves. This pattern of "entrepreneurship" or federal support has required that scientists make the basic decisions, for only they can comprehend and judge the science being funded. Political executives and the Congress simply appropriate aggregate amounts, which scientists then distribute to themselves, with *some* eye to pleasing the nation's political leadership. Social, life, engineering, and physical scientists are involved in varying degrees. Their influence, if only because it is a pork-barrel process, has been consistently high.

Science policy is formed and administered through several groups such as the National Science Foundation, Federal Council on Science and Technology,[59] White House Office of Science and Technology, President's Science Advisor, National Academy of Sciences, Bureau of the Budget (now under the Office of Management and Budget), and the President's Science Advisory Committee. Thus policy making is centered in the Executive Branch, particularly within the Executive Office of the President and the NSF. Within these centers, scientists occupy major policy and administrative positions and make key choices.

Despite the obvious fact that the subject matter and policy outcomes of the science-policy process are basically scientific, scientists' decisions on science policy are rarely aided by science. Scientific criteria are used by NSF panel members passing on individual grant applications, but, on the higher level of forming science-policy goals and objectives or allocating funds among major scientific fields, policy results from essentially political assumptions and nonscientific processes like bargaining, ideological commitments, and self-interests of specific scientific fields. Scientists make science policy, but they use their political savvy and desires more than their science.

High Levels of Scientist Influence

The scientific community may be fragmented, pluralistic, and constantly divided, but it unites on the prerogative of scientists to divide the science-policy pie. Scientists' expectations of a pork-barrel relationship with government in effect constitutes a governmental "distribution" of lump sums and scientists' "self-government" over the funds' allocation. Scientists have apparently been able to accept a hierarchy, staffed by their own kind, which will regiment and shape their behavior by selectively funding research. This would appear to conflict with scientists' norms of equality, nonregimentation, and individual freedom in scientific inquiry, but, in fact, scientific disciplines themselves are powerful hierarchies that sustain the prestige, power, and wealth accruing to scientists high in the hierarchy. Scientists' politics within their own fields probably differs little from their collective politics in making their own science policy.

Sir Solly Zuckerman verbalizes scientists' argument for controlling science, noting their special responsibility in:

> ... the process of deciding what science should be encouraged and what science applied. I say this because the scientists' knowledge of the basic facts whose application is determining the complexity of our industrial civilization is intrinsic, and not derivative. ... Looking from within at the hazy coastline which fringes the limits of scientific knowledge, he always has reason to hope that the direction into which he is peering might one day become illuminated with new understanding. Not so for the layman. When he gazes from outside upon the body of science and technology, only too often does he see reflections of his own wishful thoughts.[60]

Still, many argue that defaulting science policy-making to scientists is as dangerous as leaving military policy to generals. Such autonomy increases the probability that decisions will be based on factors internal to the scientific community and its self-interest rather than on usefulness and relevance to the society. Self-regulation might also make scientific "fads," "spectaculars," permissive granting of funds, and a pork-barrel ethic much more likely.[61]

Scientists' influence over science policy depends somewhat on a scientific "interlocking directorate" in the form of multiple committee and panel memberships. But such scientific leadership cliques do more than promote the common interests of scientists. These men, who "have climbed to power through conservative hierarchies and tend to hold conservative values," [62] tend to promote their own interests and the interests of *specific scientific fields* they represent. The leaders are predominantly physical scientists, more specifically chemists and physicists, and though they must decide on priorities involving the social science, they exhibit a noticeable bias against funds for social science and some areas within physical science.[63] These science-policy leaders are not permanent government employees but are only temporarily involved and removed from their base in universities. They are voluntary participants and face nowhere near the "command" or superior-subordinate relationships most career administrators face.[64]

Scientists making science policy are "political" scientists whose activities are more political than scientific. And science policy-making has been described by observers as being more politics than science. Scientists making science policy have minimal specialization in the economics of choice for major commitments of funds, though they are presumably adept at judging specific applications for research funds. So effective have these scientific leaders been in procuring support for scientific research that most scientists swallow their criticism in order not to disturb the cornucopia's flow.

Not all factors augment scientists' influence. Increasingly in recent years competition among scientists for funds and political leaders' reluctance to issue blank checks have been limitations on the scientist's participation and influence in science policy making.

1. COMPETITION. Daniel Greenberg has observed increased conflict among scientists over distribution of the fixed amount of science funds. Some disciplines feel that other disciplines receive more than their share and that their

own efforts are poorly funded. However, this conflict revolves around specific granting decisions, and allocative choices. It does not deal with the right of scientists' leaders to make science policy. Therefore, scientists will not willingly cede a policy-making role to nonscientists, though that may happen if science's internal conflicts require arbitration or if scientists are too preoccupied with their research.

2. DECLINE OF THE BLANK CHECK. Political leaders no longer suffer from the novelty of science in government. They are less awestruck by scientists' presence and have developed an ability to ask the "right" questions. As a result, scientists' "self-government" must show some responsiveness to political and social demands.[65]

Government has no *policy with respect to science*.[66] It has only developed a *policy for science*, which varies with the political moment and scientists' demands. Congressmen, faced with the urban crisis, Vietnam, and concern over the level of taxation, have been more willing to cut scientists' funds, except when the funded research promises some immediate application to selected social and environmental problems. Thus basic research suffers from this political preference for immediate payoffs. Scientists likewise are subjected to common norms of administrative "economy" and efficiency. Research with national security implications garners more support, and projects that might be seen as frivolous are politically vulnerable (cf. an imaginary example, "The Sex Life of the Australian Citrus Bore").

Scientists' funds originally were stimulated by fears of a missile and science gap symbolized by the Soviet Union's launching of Sputnik. But that sense of urgency has subsided and given way to other fears. As a result, Congressmen have come to perceive some science as "overfed" and "uncontrolled." However, despite budgetary cuts and threats of direct control, scientists still have effective control over what has been described as a pork barrel with the pigs themselves determining who receives the pork and which pigs get fatter than others.[67]

Notes

1. John Kenneth Galbraith, *The New Industrial State* (Boston: Houghton Mifflin Company, 1967).

2. Some sources on economists and fiscal and monetary policy making include Corinne Silverman, *The President's Economic Advisers* (University, Alabama: University of Alabama Press, 1959); Edwin G. Nourse, *Economics in the Public Service* (New York: Harcourt, Brace and World, 1953); Stephen K. Bailey, *Congress Makes a Law: The Story Behind the Employment Act of 1946* (New York: Columbia University Press, 1950); Edward S. Flash, Jr., *Economic Advice and Presidential Leadership: The Council of Economic Advisers* (New York: Columbia University Press, 1965); James L. Knipe, *The Federal Reserve and the American Dollar: Problems and Policies, 1946–64* (Chapel Hill: University of North Carolina Press, 1965); C. R. Whittlesey, "Power and Influence in the Federal Reserve System," *Economica*, XXX: 117 (February, 1963), pp. 33–44; Delbert C. Hastings and Ross M. Robertson, "The Mysterious World of the Fed," *Business Horizons*, V: 1 (Spring, 1962), pp. 97–104; and Michael D. Reagan, "The Political Structure of the Federal Reserve System," *American Political Science Review*, LV: 1 (March, 1961), pp. 64–76.

3. Flash, *op. cit.*, pp. 269, 277. Flash's analysis and use of the knowledge-power framework make his book worthwhile for anyone interested in the relationship of knowledge to power. Its contribution to this research has been sizable.

4. Herbert Stein, "Tax Cut in Camelot," *Trans-action*, 6: 5 (March, 1969), p. 39. This is a penetrating and well-written article excerpted from his book, *The Fiscal Revolution in America*.

5. *Ibid.*, p. 42.

6. Flash, *op. cit.*, p. 278.

7. *Ibid.*, p. 279.

222 High Levels of Scientist Influence

8. *Ibid.*, p. 307.

9. Whittlesey, *op. cit.*, p. 38.

10. *Ibid.*, pp. 38-40.

11. Galbraith, *op. cit.*

12. Flash, *op. cit.*, p. 314.

13. The point that foreign as well as domestic policies may have distributive, redistributive, and regulative impacts is made by James N. Rosenau (ed.), *Domestic Sources of Foreign Policy* (New York: Free Press, 1967), Chapter I; Samuel P. Huntington, *The Common Defense* (New York: Columbia University Press, 1960); Stephen J. Cimbala, "Foreign Policy as an Issue Area: A Roll Call Analysis," *American Political Science Review*, LXIII : 1 (March, 1969), pp. 148-156; and Theodore J. Lowi, "Making Democracy Safe for the World: National Politics and Foreign Policy," Rosenau (ed.), *Domestic Sources of Foreign Policy* (New York: Free Press, 1967).

14. Though none are particularly helpful, some sources on weather policy and meteorological sciences are Donald R. Whitnah, *A History of the United States Weather Bureau* (Urbana: University of Illinois Press, 1961); Wallace S. Sayre, "Scientists and American Science Policy," *Scientists and National Policy-Making*, ed. by Robert Gilpin and Christopher Wright (New York: Columbia University Press, 1964), p. 110; E. R. Piore and R. N. Kreidler, "Recent Developments in the Relationship of Government to Science," *Annals of the American Academy of Political and Social Science*, CCCXXVII (January, 1960), p. 15.

Recent treatments of weather modification raise more political issues. They include W. R. Derrick Sewell, ed., *Human Dimensions of Weather Modification* (Chicago: University of Chicago, Department of Geography, Research Paper No. 105, 1966); Robert G. Fleagle, ed., *Weather Modification: Science and Public Policy* (Seattle: University of Washington Press, 1969); Dean E. Mann, "The Yuba City Flood: A Case Study of Weather Modification," *Bulletin of the American Meteorological Society*, 49 : 7 (July, 1968), pp. 690-714; Dean E. Mann, "Public Policy Issues in Weather Modification," *Journal of the Irrigation and Drainage Division (American Society of Civil Engineers)*, 95 : IR3 (Proceedings Paper 6772, September 1969), pp. 375-384; and W. Henry Lambright, "Public Administration and Science and Technology," (Unpublished paper, 1969 Annual Meeting of the American Political Science Association, New York, September 2-6, 1969), pp. 23-25. Professor Lambright is currently writing a case history of Federal weather modification policy.

15. Analyses of scientists and health policy making are sparse. However, the reader might consult Stanley Joel Reiser, "Smoking and Health: The Congress and Causality," *Knowledge and Power*, ed. by Sanford A. Lakoff (New York: Free Press, 1966), pp. 293-311;

Wallace S. Sayre, "Scientists and American Science Policy," *Scientists and National Policy-Making*, ed. by Robert Gilpin and Christopher Wright (New York: Columbia University Press, 1964); A. Lee Fritschler, *Smoking and Politics: Policymaking and the Federal Bureaucracy* (New York: Appleton-Century-Crofts, 1969); "FDA Reveals Manipulation of Reports," *Arizona Daily Star*, May 1, 1970, p. 1; and Philip M. Boffey, "Nader's Raiders on the FDA: Science and Scientists 'Misused,'" *Science* (17 April, 1970), pp. 349–352. The results of the Nader raid on the FDA appear in a report entitled *The Chemical Feast*, edited by the task force leader, James S. Turner.

16. The field of mental health both confirms and disconfirms our premise about "life style" as a constraint. Treatment of mental ill-health is bound up with changing life styles, and many individuals accept that necessity. However, increased reliance on drugs, such as tranquilizers, indicates that even here life styles are not readily changed, and "quick technological fixes" are desired by patient and doctor alike, even though they may merely alleviate symptoms.

17. Accounts of the smoking controversy are in Reiser, "Smoking and Health," and Fritschler, *Smoking and Politics*. The labeling legislation was the result of a legislative campaign run by Surgeon General Luther Terry who, Reiser shows, used his political and scientific resources well. Fritschler's study is a comprehensive and analytically interesting effort.

18. Sayre, *loc. cit.*

19. *Science*, 160 (March 1, 1968), pp. 959–961.

20. Sources dealing with scientists in defense and deterrence policy making include Albert Wohlstetter, "Scientists, Seers and Strategy," *Foreign Affairs*, XLI : 3 (April, 1963), pp. 466–478; Wesley W. Posvar, "The Impact of Strategy Expertise on the National Security Policy of the United States," *Public Policy*, VIII (1964), pp. 36–68; Bernard Brodie, "The Strategic Scientists," *Scientists and National Policy-Making*, ed. by Robert Gilpin and Christopher Wright (New York: Columbia University Press, 1964); Albert Wohlstetter, "Strategy and the Natural Scientists," *Scientists and National Policy-Making*, ed. by Gilpin and Wright (New York: Columbia University Press, 1964); Edward Speyer, "The Brave New World for Scientists," *Dissent*, VIII : 2 (Spring, 1961), pp. 126–136; Paul Y. Hammond, *Organizing for Defense* (Princeton: Princeton University Press, 1961); Morris Janowitz, *The Professional Soldier* (New York: Free Press, 1961); Samuel P. Huntington, *The Common Defense* (New York: Columbia University Press, 1961); Fred J Cook, *The Warfare State* (New York: Macmillan Co., 1962); Joseph Kraft, *Profiles in Power* (New York: New American Library, 1966); Robert A. Levine, *The Arms Debate* (Cambridge: Harvard University Press, 1963); Arthur Herzog, *The War-Peace Establishment* (New York: Harper and Row, 1965); Charles J. Hitch and Roland N. McKean, *The Economics of Defense in the Nuclear Age* (Cambridge: Harvard

University Press, 1960); Charles J. Hitch, *Decision-Making for Defense* (Berkeley: University of California Press, 1965); W. W. Kauffman, *The McNamara Strategy* (New York: Harper and Row, 1964); John J. Clark, *The New Economics of National Defense* (New York: Random House, 1966); Warner R. Schilling, Paul Y. Hammond, and Glenn H. Snyder, *Strategy, Politics and Defense Budgets* (New York: Columbia University Press, 1962); Bernard Brodie, *The American Scientific Strategists* (Santa Monica, California: Rand Corporation, 1964); Walter Millis, Harvey C. Mansfield, and Harold Stein, *Arms and the State* (New York: Twentieth Century Fund, 1958); Harry Howe Ransom, *Can American Democracy Survive the Cold War?* (Garden City, N.Y.: Doubleday and Co., 1964); Julius Duscha, *Arms, Money and Politics* New York: Ives Washburn, Inc., 1965); Bruce L. R. Smith, *The Rand Corporation* (Cambridge: Harvard University Press, 1966); Ralph E. Lapp, *The New Priesthood* (New York: Harper and Row, 1965); and Raymond V. Bowers, "The Uses of Sociology in the Military Establishment," *The Uses of Sociology*, ed. by Paul Lazarsfeld et al (New York: Basic Books, 1967), pp. 234–274.

21. The best sources for information on the decision to use the atomic bomb are Len Giovannitti and Fred Freed, *The Decision to Drop the Bomb* (New York: Coward-McCann, 1965); Joan W. Moore and Burton M. Moore, "The Role of the Scientific Elite in the Decision to Use the Atomic Bomb," *Social Problems*, VI : 1 (Summer, 1958), pp. 78–85; Henry L. Stimson, "The Decision to Use the Atomic Bomb," *The Atomic Age*, ed. by Morton Grodzins and Eugene Rabinowitch (New York: Basic Books, Inc., 1963); and Herbert Feis, *Japan Subdued* (Princeton: Princeton University Press, 1961).

22. The shelter and fallout controversies are particularly instructive on the problem of "inexpert" advice. Note Willard Libby's own shelter succumbing to a Southern California brush fire and Edward Teller, Willard Libby, and Linus Pauling's questionable use of statistics and "scientific data" in behalf of a fallout viewpoint in *Life* magazine. The shelter and fallout debates are discussed in James L. McCamy, *Science and Public Administration* (Birmingham: University of Alabama Press, 1960), pp. 189–191. Scientists' positions on various weapons systems are discussed in the following section on "weapons policy."

23. The original source on "scientific strategists" is Bernard Brodie, *The American Scientific Strategists*. However, the reader might also consult Brodie, "The Scientific Strategists," *Scientists and National Policy-Making*; Posvar, "The Impact of Strategy Expertise"; Wohlstetter, "Scientists, Seers and Strategy"; and Wohlstetter, "Strategy and the Natural Scientists," *Scientists and National Policy-Making*.

24. Lapp, *op. cit.*, p. 110.

25. Janowitz, *op. cit.*, pp. 9–10.

26. Brodie, "The Scientific Strategists," *Scientists and National Policy-Making*, p. 253.

27. *Ibid.*

28. *Ibid.*, p. 244.

29. The nature, assumptions, and political effects of such various techniques are discussed in Aaron Wildavsky, "The Political Economy of Efficiency: Cost-Benefit Analysis, Systems Analysis, and Program Budgeting," *Political Science and Public Policy*, ed. by Austin Ranney (Chicago: Markham Publishing Co., 1968), pp. 55–82. The reader might also consult Virginia Held, "PPBS Comes to Washington," *The Public Interest*, 4 (Summer, 1966), pp. 102–115; William Gorham, Elizabeth Drew, and Aaron Wildavsky, "Symposium on PPBS: Its Scope and Limits," *The Public Interest*, 8 (Summer, 1967), pp. 4–48, and "Symposium: PPBS Reexamined," *Public Administration Review*, XXIX : 2 (March–April, 1969), pp. 111–202. Several authors in the latter volume comment on the limitations and problems encountered with PPBS.

The fact that PPBS has not transferred easily from the Department of Defense to other Departments may be due to a lack of trained personnel in those areas as well as lower research and analysis budget capabilities.

Furthermore, as Allen Schick argues in his contribution, PPBS may meet resistance in any organization because it threatens to replace older decision-making patterns. For Schick, PPBS involves "systems politics" (comprehensive wholistic policy making, recognition of interdependence) rather than "process politics" (incremental, nongoaloriented, focus on bargaining and activity of politics). Thus PPBS has a built-in bias that favors certain objectives and officials:

"Process politics (and budgeting), therefore, tends to favor the partisans such as agencies, bureaus, and interest groups, while system politics (and budgeting) tends to favor the central allocators, especially the chief executive and the budget agency. System politics also can be used to bolster certain officials who have mixed mobilizing-rationing roles." (p. 139)

And so many officials in an organization might well resist the implementation of such a new technique or way of doing things. Some would benefit in increased power and respect at the expense of individuals currently in leading positions. See Allen Schick, "Systems Politics and Systems Budgeting," *Public Administration Review*, XXIX : 2 (March–April, 1969), pp. 111–202.

Systems analysis, policy sciences, PPBS and allied policy-making techniques may (1) place power in newly-hired hands; (2) force cooperation and independence; (3) de-emphasize high risk actions; (4) raise more policy alternatives to respectability; and (5) find basic values and political assumptions difficult to quantify or study systematically.

I have argued this point elsewhere, that some "behavioral technologies" such as PPBS are less politically acceptable because they may involve a redistribution of power, wealth, and respect among men. Such may be the case with the "bias" Schick identifies in PPBS.

See my "Political Arenas and the Contributions of Physical and Behavioral Sciences and Technologies to Policy-Making" (Paper prepared for the 1969 Meeting of the American Political Science Association, Commodore Hotel, New York, September 2–6, 1969), revised as "Political Arenas, Life Styles, and the Impact of Technologies on Policymaking," *Policy Sciences*, I : 3 (tentative, Fall, 1970).

30. The strategic bases study and its implementation are well analyzed in Bruce L. R. Smith, *op. cit.*, pp. 195–240. Smith contends that built-in safeguards apparently prevent "arbitrary action based on uncritical deference to the scientific adviser." (p. 238) Yet the impact of the study does indicate significant scientist influence, especially when that research is done by men working outside the bureaucracy where they can be free from pressures for symbolic and supportive "white-washing" research. See Joseph W. Eaton, "Symbolic and Substantive Evaluative Research," *Administrative Science Quarterly*, VI (March, 1962), pp. 421–442.

31. Wohlstetter, "Scientists, Seers and Strategy."

32. Samuel P. Huntington, *op. cit.*, pp. 123–196, 284–368.

33. *Ibid.*, p. 288.

34. Smith, *op. cit.*, p. 239.

35. Sources on scientists and weapons policy making include David W. Tarr, "Military Technology and the Policy Process," *Western Political Quarterly*, XVIII : 1 (March, 1965), pp. 135–148; Joan W. Moore and Burton M. Moore, "The Role of the Scientific Elite in the Decision to Use the Atomic Bomb," *Social Problems*, VI : 1 (Summer, 1958), pp. 78–85; James Reston, "The H-Bomb Decision," *New York Times*, April 8, 1954, p. 20; Warner Schilling, "The H-Bomb Decision: How to Decide Without Actually Choosing," *Political Science Quarterly*, LXXVI : 1 March, 1961), pp. 24–46; "The Hidden Struggle for the H-Bomb," *Fortune*, XLVII (May, 1953), pp. 109–110, 230; Henry L. Stimson, "The Decision to Use the Atomic Bomb," *The Atomic Age*, ed. by Morton Grodzins and Eugene Rabinowitch (New York: Basic Books, 1963); Herbert Feis, *Japan Subdued* (Princeton: Princeton University Press, 1961); Merton J. Peck and Frederic M. Scherer, *The Weapons Acquisition Process: An Economic Analysis* (Boston: Harvard University Graduate School of Business Administration, 1962); Robert J. Art, *The TFX Decision* (Boston: Little, Brown and Company, 1968); Fred J. Cook, *The Warfare State* (New York: Macmillan Company, 1962); Samuel P. Huntington, *The Common Defense* (New York: Columbia University Press, 1961); and Andrew Hamilton, "High Flying in the Pentagon," *New Republic*, 160 : 22 (May 31, 1969), pp. 16–18. Works cited in the discussions of defense and arms-control policy making might also be useful.

36. *New York Times*, December 27, 1967, p. 8.

37. The major sources dealing with the development and use of the atomic bomb include Stephane Groueff, *Manhattan Project* (Boston: Little, Brown and Co., 1967); Michael Amrine, *The Great Decision* (New York: G. P. Putnam's Sons, 1959); George C. Batchelder, *The Irreversible Decision* (New York: Macmillan Company, 1961); Moore and Moore, *op. cit.;* Stimson, *op. cit.;* and Feis, *op. cit.*

38. Robert Jungk, *Brighter Than a Thousand Suns* (London: Victor Gollancz, Ltd., 1958). Although many observers dismiss his argument as fantasy, Jungk suggests that the American scientists should have withheld or delayed development of the bomb or at least tried to make an agreement with the German scientists who, he argues, were delaying their own work. His criticism is important for this research not because I am interested in making moral judgments, but because scientists are treated as behaving, responsible actors whose actions may shape public policy.

39. Lewis S. Feuer, *The Scientific Intellectual* (New York: Basic Books, Inc., 1963), p. 395.

40. Stanley Jarolem and Daniel Kahn, "TFX Trouble," *New Republic*, April 27, 1968, pp. 18–21.

41. Art, *op. cit.*

42. Project Hindsight was a Defense Department study of the usefulness of research from World War II through 1962. The study, clearly with methodological bias, did contend that basic research was responsible for very few research developments or "events" in histories of weapons systems. See C. W. Sherwin and R. S. Isenson, "Project Hindsight," *Science*, CLVI, June 23, 1967, pp. 1571–1577.

43. Peck and Scherer, *op. cit.*, p. 237.

44. Ralph E. Lapp, *The Weapons Culture* (New York: W. W. Norton and Co., 1968). See other discussions of the "military-industrial complex" in Arthur Herzog, *The War-Peace Establishment* (New York: Harper and Row, 1965); Harold P. Green and Alan Rosenthal, *Government of the Atom* (New York: Atherton Press, 1963); and Cook, *op. cit.*

45. Peck and Scherer, *op. cit.* Furthermore, an Arthur D. Little study once revealed a long time-lag between initial discovery of feasibility or design and eventual application in five years. See U.S. Congress, House, Research and Technical Programs Subcommittee of the Committee on Government Operations, *The Use of Social Research in Federal Domestic Programs*, 90th Cong., 1st Sess., 1967, pp. 485–486.

46. Tarr, *op. cit.*, pp. 140–141.

47. *Ibid.*, p. 140. The insight in the Tarr article and his characterization of weapons policy making is a valuable contribution worth reading. His comments on the divorce of military instruments from

policy objectives, the tendency of technology to determine policy, the difficulties in relating means to ends, and the problems posed by secrecy and technical complexity are most perceptive.

48. Don K. Price, *The Scientific Estate* (Cambridge: Harvard University Press, 1965), p. 151.

49. Sources on the hydrogen bomb controversy include Huntington, *op. cit.;* Schilling, *op. cit.;* Reston, *op. cit.;* Robert Gilpin, *American Scientists and Nuclear Weapons Policy* (Princeton: Princeton University Press, 1962); Stefan J. Dupre and Sanford A. Lakoff, *Science and the Nation* (Englewood Cliffs: Prentice-Hall, 1962); Warner Schilling, "Scientists, Foreign Policy, and Politics," *Scientists and National Policy-Making* (New York: Columbia University Press, 1964); and "The Hidden Struggle for the H-Bomb," *Fortune,* XLVII (May, 1953), pp. 109–110, 230.

50. The most perceptive account of Truman's decision is Schilling, *op. cit.*

51. The best material on scientists' efforts to counteract what they saw as an overreliance on super-bombs is in Robert Gilpin, *American Scientists and Nuclear Weapons Policy,* pp. 115-121, 219, 288ff.

52. Robert H. Salisbury and John P. Heinz, "A Theory of Policy Analysis and Some Preliminary Applications," *Policy Analysis in Political Science,* ed. by Ira Sharkansky (Chicago: Markham Publishing Co., 1970), pp. 39–60.

53. Sources on scientists and space policy include Gordon J. F. MacDonald, "Science and Space Policy: How Does It Get Planned?" *Bulletin of the Atomic Scientists,* XXIII : 5 (May, 1967), pp. 2–9; Alison Griffith, *The National Aeronautics and Space Act* (Washington: Public Affairs Press, 1962); Lillian Levy (ed.), *Space: Its Impact on Man and Society* (New York: W. W. Norton and Co., 1965); Vernon Van Dyke, *Pride and Power: The Rationale of the Space Program* (Urbana: University of Illinois Press, 1964); Amitai Etzioni, *The Moon-Doggle* (New York: Doubleday and Company, Inc., 1964); Donald W. Cox, *America's New Policy Makers* (Philadelphia: Chilton Co., 1964); and H. L. Nieburg, *In the Name of Science* (Chicago: Quadrangle Books, 1966).

54. Jules Verne, *A Trip from the Earth to the Moon* (New York: Vincent Parke and Company, 1911). The quotation appears on the introductory page to Chapter VI on "The Permissive Limits of Ignorance and Belief in the United States."

55. Cox, *op. cit.,* pp. 106–110.

56. Etzioni, *op. cit.,* p. 11.

57. *Ibid.,* pp. xiii–xiv.

58. Sources on scientists and science policy making include Emmanuel G. Mesthene, "Can Only Scientists Make Government Science Policy?" *Science*, CXLV (July 17, 1964), pp. 237–240; Alvin M. Weinberg, "Criteria for Scientific Choice," *Science and Society*, ed. by Norman Kaplan (Chicago: Rand McNally, 1965); Daniel S. Greenberg, "When Pure Science Meets Pure Politics," *Reporter*, XXX : 6 (March 12, 1964), pp. 39–41; Wallace S. Sayre, "Scientists and American Science Policy," *Science*, CXXXIII (March 24, 1961), pp. 859–864; Daniel S. Greenberg, "Mohole: The Project That Went Awry," *Knowledge and Power*, ed. by Sanford Lakoff (New York: Free Press, 1966); James L. Penick, Jr. (ed.), *The Politics of American Science* (Chicago: Rand McNally, 1965); Daniel S. Greenberg, *The Politics of Pure Science* (New York: New American Library, 1967); Norman W. Storer, "Some Sociological Aspects of Federal Science Policy," *American Behavioral Scientist*, VI : 4 (December, 1962), pp. 27–30; Stephen Toulmin, "The Complexity of Scientific Choice: A Stocktaking," *Minerva*, II : 3 (Spring, 1964), pp. 343–359; Michael D. Reagan, *Science and the Federal Patron* (New York: Oxford University Press, 1969); and H. L. Nieburg, *In the Name of Science* (Chicago: Quadrangle Books, 1966).

59. The Federal Council on Science and Technology is a cabinet-level group of "science" related personnel which acts as a clearing house and coordinator for federal science activities.

60. Sir Solly Zuckerman, *Scientists and War* (London: Hamish Hamilton, 1966), p. xi.

61. Project Mohole is an example, though perhaps an extreme caricature, of the results of self-government. A group of scientists, dismayed with a series of unimaginative proposals presented to their panel for review, decided to back one of their own—an impulsive, ambitious desire to drill deep into the earth's Mohorovicic discontinuity from an ocean platform. The whole affair, ending in Congressional pressures for cancellation, takes on an absurd air. Daniel Greenberg's excellent account is worth reading. Greenberg, "Mohole: The Project That Went Awry," *Knowledge and Power*, ed. by Sanford A. Lakoff (New York: The Free Press, 1966), pp. 87–111.

62. Center for the Study of Democratic Institutions, *Science, Scientists, and Politics* (New York: Fund for the Republic, 1963). See the comment by Donald W. Michael, p. 7.

63. The bias against social and behavioral sciences stems from (1) their omission from NSF's original mandate and (2) physical scientists' feeling that they are not "scientific" and fear that their greater political relevance might irk politicians and render support for all science more vulnerable.

64. Robert C. Wood, "Scientists and Politics: The Rise of an Apolitical Elite," *Scientists and National Policy Making*," ed. by Robert

Gilpin and Christopher Wright (New York: Columbia University Press, 1964).

65. Greenberg, *The Politics of Pure Science*, pp. 270–293. Greenberg's analysis of a shift from old to new politics of science is perceptive, but we differ over the extent of the shift.

Government's early support for science in the United States centered around such useful fields as the Coast Survey, exploration of Western lands, agriculture, nautical technologies, meteorology, the Geological Survey, and mining through the Bureau of Mines. Utilization was paramount in importance as a justification for support. Political considerations even were involved, as in the northwestern states' pressures for agricultural research and development and the whole nation's defense and war-related needs. Cf. A. Hunter Dupree, *Science in the Federal Government: A History of Policies and Activities to 1940* (Cambridge: Harvard University Press, 1957).

Since 1850 government's support for basic research has been noticeable, though not always readily apparent. After World War II, when Vannevar Bush was urging massive support for science and basic research to conquer that "last frontier," government began a deliberate effort for science. Sputnik accelerated that effort immeasurably. NSF's coffers became a source of much funding. Government increasingly *contracted research out* to nonprofit organizations (RAND, ANSER, Systems Development Corporation, MITRE, etc.) and private industries (Union Carbide at Oak Ridge, GE at Hanford, Dupont at Savannah River, Pan Am at Cape Kennedy, etc.). The "contract state" had arrived.

The middle 1960s saw increased ferment in science policy making. The policy-making game became more "zero sum" as budgetary cuts stimulated by inflation and the Vietnam war either reduced or held constant the funds available for scientific research. The project system gave way to grants to institutions. Scientists would thus more often have to petition to their university rather than to Washington administrators for funds. Prior to the project system, they would appeal for funds to a panel of their peers. So in a sense decisions on individual proposals have been removed from the hands of scientists' peers and given to scientists who serve as either government or university administrators.

With fewer funds available, pressures increased from the political sector for "more relevant" research. Funds would go to scientists whose research promised solutions and benefits to certain key areas like urban problems, transportation, states' problems, fuels, molecular biology and genetics, medicine, and marine resources. Studies of policy alternatives were encouraged. There was concern with the "transfer" of technologies to civilian uses. Many others expressed a desire to assess the impact of new technologies prior to their introduction, as society might well have done with gasoline engines (pollution) and supersonic travel (jet noise). And the rise of social problems on the domestic scene brought on increased funds for social scientists. In 1970 NSF organized a new Program of Interdisciplinary Research Relevant to the Problems of Our Society (IRRPOS).

No longer did there seem to be sufficient funds to go around. Choices had to be made, and this required consideration of "criteria." Was a field ready for development and breakthroughs? Were existing scientists in a field competent? What technologies might be produced? Did the research promise benefits to man's welfare and values? Could the research be considered a capital investment in future discoveries and uses? Could we support science for the same reasons we support the creative arts? Did the research promise "spillover" benefits for other scientific fields? Did research offer benefits for national defense or more domestic freedom? Did the funds benefit one geographical area unduly, or should funds be spread out over various geographical regions and states?

66. Storer, *op cit.*, pp. 27–30. Michael D. Reagan sees what he calls a *de facto* science policy. Such a policy results from a series of decisions and activities in which assumptions, guidelines, partial policies, and priorities indicate an overall "policy." Such policy is implicit and unstated. Nevertheless this kind of policy is the policy I am describing here. Reagan, *op. cit.*, pp. 106–107.

67. Toulmin, *op. cit.*, pp. 350ff. The initiator of demands has the advantage in a pork-barrel situation. Science policy tends therefore to be inclusive rather than make "yes" or "no" choices.

Science, Scientists, and National Policy Making

CHAPTER 7

Scientists' influence depends first upon their performance. They must discover scientific facts and develop scientific knowledge. Without these contributions, the scientist has no scientific findings to offer the policy-maker. Without the foundation of science, the scientist could have no scientific *reputation* to use in influencing policy, whether he would speak from personal research or wander beyond his own competence and scientific judgment into "political" matters.

Summary: Scientists' Influence and the Context of Public Policy Making

Several generalizations emerge from the analysis of scientists' influence in Chapters 4, 5, and 6. As generalizations, they can serve simultaneously as a *summary of the analysis* and tentatively set forth some interesting *relationships for further research*. Influence, then, might be

NOTES TO CHAPTER 7 START ON PAGE 253

viewed as related to seven conditions surrounding the scientists' involvement in policy making.

1. *Scientists' influence within the Executive Branch increases in non-zero sum policy situations and decreases in zero sum policy situations.* Zero sum situations occur when some groups or individuals gain what others lose and at others' expense. Non-zero sum situations occur when all either gain or lose together, or when some appear to gain or lose without affecting others' resources.

Distributive, communal security, and entrepreneurial arenas harbor less domestic conflict. That is, significant groups in the system do not see the impact of government activity as divisive. The groups tend to see everyone benefiting from policy in these arenas and do not consider one another as rivals for a pie of fixed size. Redistributive and self-regulative arenas, on the other hand, have impacts perceived differently by significant groups. Values, whether wealth, skill, respect, enlightenment, well-being, or security, are threatened with redistribution. Major groups see one another as rivals. Such rival groups may be government and extractive industries in conflict over government's role in extraction policy-making or "haves" and "have-nots" in conflict over the division of national wealth. Regulative and extra-national arenas also evoke rivalries, though of a different type. Regulation involves a conflict pitting government, allied with the "public," against a given sector or sectors in the society. Extra-national policies often threaten the "redistribution" of values from Americans to non-Americans.

Scientists are needed by the U.S. Executive Branch in non-zero sum situations to provide resources or values for distribution. These values may include military security, transportation technologies, space exploration, weapons, economic stability and security, health and physical well-being, agricultural knowledge, and funds for scientific research. Furthermore, these distributed values and policies enjoy broad support either as "pork barrel" politics or as

Science, Scientists, and National Policy Making

contributions to the communal interest. Therefore, the Executive Branch runs no risk in using scientists, and it benefits insofar as the scientists increase the amount and effectiveness of values the government is distributing.

Scientists influencing fiscal and monetary policy contribute to a distribution of economic security and well-being in an economic management arena and fiscal and monetary policy. Scientists in agriculture policy making and transportation policy making enable the government to distribute better and more benefits to special groups and the general interest. In the realm of health and weapons policy making, scientists' influence stems from increased levels of communal security scientists contribute. In the instance of entrepreneurial situations in which scientists are among the beneficiaries of government's distribution of values, their influence again will be felt. Space policy and science policy illustrate this latter situation.

When groups and leaders see the impact of policy as undesirable or not in their interest, then the Executive Branch must move cautiously and often is reluctant to move at all. Science then is less needed and may not play as critical a role. Scientists whose fields are potentially relevant to social policy making must face this barrier to influence. Governmental organization evokes a similar situation regarding the use and influence of scientists because power may be threatened with redistribution. The entire extra-national policy arena raises the spector of "redistribution" to non-Americans at the expense of Americans. Foreign aid policy making has been bound by similar constraints, and scientists' influence has been restricted. Scientists' influence over disarmament and arms control policy and foreign political policy has also been diminished by the "redistribution" associated with policies in these fields.

However, should government gamble and act in a redistributive manner (e.g., progressive taxation, bureau reorganization, foreign aid, welfare payments, low-income housing), then scientists may be needed to reduce uncertainty, buttress the government's argument or render its action more effective and less offensive to groups being harmed. Or scientific reasons and scientists' recommendations may be used to justify a redistributive action, removing the decision from "politics." If the contending groups are deeply polarized, then science may be the sole acceptable means to resolving conflict and acting. Scientists then provide arbitration rather than mere advice. But all groups must agree to let scientists decide —a mutual agreement most unlikely in American politics, where citizens and leaders prefer *political* solutions.

Redistribution may be made less "offensive" by finding impersonal technological means to redistribution of values, manipulating the policy debate to make the proposed course seem less "redistributive," or through government merely changing the rules and payoffs governing groups' competition rather than being the actual transfer agent for the redistribution.

2. *Scientists' influence within the Executive Branch increases when the Executive Branch, acting in its own or the public's interest, becomes an actor or combatant in the society and seeks to deprive or allocate values among major sectors in the society.* Such influence-creating situations occur in regulative and redistributive arenas. Scientists' influence rises when the executive assumes an active role in social policy, antitrust policy, transportation safety policy, conservation policy, pollution policy, and trade and balance-of-payments policy. Influence here is moderate in such "regulative" arenas.

Government is more likely to act in a regulative manner than in a redistributive manner, because opposition is weaker if government is perceived to merely deprive one sector (regulation) rather than shift values from that sector to another (redistribution). However, in both cases, because its actions are politically sensitive, the Executive Branch will need effective strategies, accurate diagnoses of the situation to be regulated or redistributed, and justification for regulative or redistributive action. Scientists provide these resources, and the government needs them because it is an active combatant in a "zero sum game."

Government's action, for scientists to be needed, must involve "regulation" of a specific sector and not general public behavior. Regulation is politically attractive because it strikes at a specific and somewhat divided sector. Therefore, a government would be reluctant to "regulate" the lives of its general citizenry, particularly if that regulation meant changing citizens' traditional patterns of behavior and life styles (cf. pollution, health, weapons policy). If scientists' contributions related to these general interests rather than special interests, then scientists are not likely to be needed by government.

More often than not, however, government refrains from regulative and redistributive activities. There are less controversial and equally politically productive courses of action such as distributive and communal security activities. But when government does regulate or redistribute, choosing not to reaffirm the status quo and benefits accruing to dominant interests, scientists may be needed and should be that much more influential.

3. *Scientists' influence within the Executive Branch decreases insofar as the impact or expected impact of a policy involves procedural rather than substantive change.* When the government seeks to change the "rules" governing policy making (cf. poll tax, regulatory procedures, active or inactive governmental role), it affects and arouses much more fundamental political interests. Thus scientists' influence in extraction policy-making is low in part due to the fact that extraction industries see a more active government role as a change in the "rules." The "rule" change here would be a shift from *self-regulation* toward *regulation*. Scientists would give government the knowledge and capacity to "regulate" their industry actively or even become a competing entrepreneur. This would involve a procedural rather than substantive change and much opposition.

But government itself is often the object of procedural changes. Organization policy-making deals with this type of policy activity, and even here scientists are not needed when their contribution goes beyond "efficiency" and "economy." As a rule, no government will easily move to internally redistribute political power within the councils of government. It may, however, more readily encourage scientists to develop what Yehezkel Dror calls "metapolicymaking." [1] This would include such scientific skills as personnel development, policy sciences, decision theory, game theory, systems analysis, cost-effectiveness techniques, operations research, and communications analysis.

However, these sciences' contributions will be less influential the more they involve a change in existing power distributions within the Executive Branch. Furthermore, it is probably more difficult to alter policy-makers' behavior in making policy than to alter their substantive views on the content of that policy.

4. *Scientists' influence within the Executive Branch increases when the decision system is integrated, needs scientists, and agrees with the implications of the scientists' contributions. When the decision system or Executive Branch is fragmented or divided, scientists' influence may decline. When the formal or governmental decision system is integrated and nongovernmental demands are fragmented, then the executive is more likely to act and, in acting, need scientists.*[2] The Executive Branch, when it is a fragmented decision system, is an inhospitable environment for scientists' participation and influence. Such fragmentation and lack of integration may stem from divisions among policy-makers themselves or contamination from divisions in the society. Inability to act may result from policy-makers' failure to agree on the desirability of a course of action. In either case, neither climate nor inaction will be conducive to scientists and scientists' influence.

General support from an integrated system of demands

usually leads to the use of scientists to achieve the commonly demanded objectives. Therefore, scientists in the communal security arena and economic management arena become involved and influential (cf. fiscal and monetary policy, weather policy, health policy, deterrence and defense policy, and weapons policy). General united opposition, conversely, would stifle executive action and diminish any reliance on scientists.[3]

Support or opposition, it is argued, hinge on the extent to which important values are threatened or offered. Therefore, an aura of support surrounds governmental activities in the distributive, communal security, economic management, and entrepreneurial arenas because desired values are commonly served and none threatened. When this occurs, the need for science and scientists is heightened. Scientists are better funded, develop more useful and competent science, occupy policy-making positions, and contribute essential advice. Then, from the standpoint of scientists' influence, such situations are most conducive.

Sometimes, however, significant vested interests, benefiting from their own political strength and an apathetic or unaware public, can fend off government regulatory action. Scientists' influence, as in extraction policy making will be adversely affected insofar as it negatively affects the vested interest. Similarly, scientists may not be needed or wanted by the Executive Branch if politically significant groups want the status quo maintained. And finally, the executive will not need scientists and their influence if there are no evident constituencies to reward or tangible demands to be met.

Regulatory activities are made possible by a fragmented system of demands. One sector of a society may be protected by regulation. Demands are not integrated. The "public" wants polluting, unsafe, monopolistic, and wasteful activities regulated. Some industries want their competitors' behavior regulated. The potentially "regulated" do not want regulation. But the groups' fragmentation leaves some room and support for government to act. This situation, conducive to scientists' use and influence, occurs in the regulatory arena and antitrust, pollution, transportation safety, trade and balance-of-payments, and conservation policy making.

5. *Scientists' influence within the Executive Branch decreases to the extent that policy involves "symbolic" rather than "tangible" activities.*[4] Science, with the exception of some social sciences, deals in tangible products. When government acts with symbols or symbolic actions, as it does in many of the policies studied, only the psychologist, political technician, social scientist, or "scientific strategist" may be useful. Usually, policy-makers perceive themselves already to have skills in symbol manipulation and symbolic action. Scientists then are less in demand. Government's actions in self-regulative, social redistributive, regulative, and governmental redistributive arenas are often more symbolic than tangible. And so scientists are that much less useful and influential in the formation of organization, social, extraction, and even some forms of regulatory policy.

6. *Scientists' influence within the Executive Branch increases when available or potential means to public policy objectives are material or "technological" rather than behavioral.* Physical scientists, building on past and new knowledge, develop material technologies that enable society to attempt solutions to its problems without drastically altering individuals' preferences and life styles. There may be dollar costs involved, but styles of living need not change.

Contraceptive devices, auto exhaust emission control devices, and weapons systems are such means-to-ends that do not require changed patterns of living. Seat belts may be effective but unacceptable styles of living in an automobile, but rapidly inflating air bags hidden in the dash may free a man from being "tied down" and protect him only on impact, when he needs protection.

So, too, does man have a passion for technological means rather than human commitments for waging war. Many weapons are impersonal, and many have the added advan-

Science, Scientists, and National Policy Making

tage of supplanting the foot soldier. Americans might have been willing to wage Vietnam-style wars with technology but not with human lives and time, a preference particularly crucial when the conflict appeared more and more as a "redistribution" of values from the United States to both Vietnams rather than a protection of Americans' communal security. In this sense, Vietnam has involved a transition from concern over foreign defense to concern for domestic impacts and considerations.

Though these "quick technological fixes" often treat symptoms of problems (e.g., distribution of air conditioners in Watts) rather than causes themselves, they are more politically and personally attractive.[5] They make the task of a regulator and provider of communal security much easier, inasmuch as measures he proposes evoke less social and political resistance.

Social scientists, on the other hand, propose solutions to problems that require changes in human attitudes, behavior, and life style (cf. social policy making). Therefore, though their proposed solutions may no doubt be correct and appropriate, there is understandable reluctance among policy-makers to pursue such a course.

Scientific professions, such as medicine, straddle a similar dilemma. Doctors prescribe technological means to health (pills, injections), and they are reasonably successful. But when their solutions depend on changed behavior (low starch diets, exercise patterns, no more smoking, intensive self-analysis), they meet with less success. Health policy-making illustrates this quite well.

Scientists' forte lies in technological means. They are therefore useful to the policy-maker. This does not rule out contributions by scientists that would imply behavioral means to change, but it does mean that policy-makers have a preference for the technological "fix."

Scientists will also be more influential to the extent that they can declare a certain course or policy "impossible" or "infeasible." Witness scientists' arguments on the nuclear-powered aircraft (ANP) project. Experts argued that no foreseen material or technological device was light enough to make such a project feasible. Furthermore, economists have cast doubts on the economic payoffs and sonic boom engineers raised questions about the noise factor for the proposed supersonic transport (cf. transportation policy). Physical scientists can rather more effectively make the physical "impossibility" argument than social scientists can make the behavioral "impossibility" argument. People see behavior as changeable and physical laws as immutable, though in fact it may be easier to change "things" than people. And social scientists, too, are perceived to speak in less precise, less certain, and less competent terms.

7. *Scientists' influence within the Executive Branch increases when scientists' own vested interests are threatened or potentially rewarded.* Science policy making and space policy making both provide some degree of verification for this generalization. Scientists in both cases benefit from governmental entrepreneurship and distribution in these fields. They, like any other group whose interests are intertwined with government and benefited by government, will secure some measure of access to policy-makers, positions of policy responsibility, and consequent additional influence. There is also basis for assuming that research funds and prestige may follow from giving policy advice. Relevance or access may be used to create a vested interest, and that potential interest might shape scientists' behavior in influencing policy.

Implications for the Study of Public Policy

Scientists' influence within the Executive Branch depends to a significant extent on the policy arena and "impacts" associated with the arena. Politically significant groups, comprising a pattern of demands, perceive governmental activities associated with a certain policy to have particular impacts.[6]

Possible impacts include the distribution and redistribution of values, such as wealth, skill, respect, well-being, power, and enlightenment. Impacts may include the regulation of values accruing to special interests or sectors in society or involve the distribution of communal security to the general citizenry. They may include government acting as an entrepreneur or remaining aloof to allow "self-regulation."

These impacts affect the use and influence of scientists through (1) their effect on governmental activities generally and (2) the policy-makers' perception of the reception politically significant groups are going to give to certain activities. Scientists are needed either when government confronts a hostile interest or decides to fulfill the significant groups' demands. And scientists are needed for their skills, participation in policy making, and influence directed toward more "effective" policy.

In sum, political factors (demands coming from politically significant groups) shape scientists' participation and influence within the Executive Branch. This finding contributes to an on-going and more general discussion among students of public policy making. Some have argued that *political factors* (interparty competition, demands, forms of government, policy-makers' attitudes, etc.) best explain policy outcomes and outputs.[7] Others contend that *socio-economic factors* (income, urbanization, industrialization, tax base, etc.) explain more variations in policies.[8]

I would argue that the influence and participation of scientists in policy making depends on *both* political and socio-economic factors though only political factors have been the subject of this research. I would further argue that political factors may in fact shape the differential impact of various sciences on public policy by shaping the amount of resources diverted to each science.

Society and government's *total* use of science and scientists surely varies with the socio-economic system's ability

244   Science, Scientists, and National Policy Making

to support science. More wealthy and industrialized systems have more resources to distribute to scientific research and scientists. Thus system resources explain a society's aggregate capacity to fund and use science.

But though all sciences may benefit from a richer society, some benefit more than others. Political factors, like patterns of demand and politically significant groups anticipating "impacts" of policies, may shape these differential levels of support and influence for specific types of scientists. Some sciences and scientists are supported more heavily than others. Some are needed and have opportunities to influence policy-makers more than others. These differences are partly due to the differing "impacts" various policies are perceived to have on politically significant groups. Policy arenas, defined in terms of impacts, harbor implications for the use, nonuse, and influence of particular fields of scientists.

So political factors more than socio-economic factors may shape the society's commitment of resources to science versus other sectors and to one branch of science versus another branch. These political factors can explain why society divides its pie between the cities and science or between physics and sociology.

Some sciences are heavily supported because the knowledge, technology, and advice they produce contributes toward the protection or production of important values. These values are sufficiently valued that citizens, politically significant groups, and policy-makers feel that scientists' contributions must be as "competent" as possible. System resources are then diverted and emphasized for research in these specific fields.

Thus scientists whose knowledge, technology and advice contribute to the protection and production of physical security, enlightenment, power, skill and wealth are more heavily endowed than scientists related to values such as respect, affection, and rectitude. Society tends to see its insecurities reduced through more wealth, power, technology, and military or police strength than through the distribution or redistribution of such democratic values as respect, affection, and rectitude. Scientists promoting or protecting the former values are thus more likely to be needed, influential, and occupy administrative, prescriptive, and policy-making positions within the Executive Branch.

Often important values are so threatened that policy-makers will, in the short run, turn to "incompetent" and undeveloped sciences and scientists for knowledge, technology, and advice. Policy-makers have no other choice. Atomic energy began in such fashion, and the social sciences have lately received a boost from urban problems, poverty, and rioting. However, in the latter case, the government's response and perception may be as much economic and military as social and scientifically based.

Finally, the social scientist and physical scientists have slightly different images in policy-makers' and citizens' minds. Physical scientists are guarantors of material progress and security through their research. Social scientists, though seen as a source of a society's "self-consciousness," are nevertheless felt to be producers and educators of "good" citizens. The research and knowledge function is de-emphasized. This conception is not unrelated to policy-makers' evaluation of social scientists' competence, status as scientists, and role in policy formation.

One important factor remains. Scientists' influence is affected by the perceived impact of activities (political factor) and by the allocation of resources to their science's development. These are the bases of their influence. But on that foundation, the scientists' own desire to influence policy and pursue power emerges. And their desire for influence and power has no doubt increased their impact on policy and occupancy of formal policy-making positions. Science and space policy making are the more obvious examples, but pollution, disarmament, and defense policy also show this pattern.

However, many scientists may desire influence over policy and policy-making positions but are frustrated by a lack of research support, reputation for scientific competence, need by policy-makers, and threatened or desired values they can protect or promote. The public, politically significant groups, and policy-makers in an arena do not see these scientists as relevant. To some extent this is the dilemma of the social scientist wanting to influence policy. There are rather severe limits on his efforts, limits which cannot be significantly altered in the immediate future.

Implications for Policy Making and Recommendations for Policy-Makers

An inquiry into scientists' participation in public policy making must have some "lessons" for policy-makers. What are the implications for participants in policy making? What problems or opportunities need their attention? What might we do differently? What have we not recognized?

Scientists augment policy-makers' and policy making's capabilities and capacities for acting effectively. They provide reasonably *distortion-free descriptions and diagnoses of reality*. Policy-makers, thus having "true" pictures of their environment, will behave more rationally. But scientists also enable policy-makers to *anticipate and predict the future of social relationships, politics, and technological developments.*

Scientists *provide justifications, some ideological and some scientific, for various policies and activities. Alternative courses of action and strategies are multiplied. Resources for policy formation and action are strengthened,* just as political advisers and legal advice strengthen policy making.

Technologies, arising out of science and developed by scientists, give policy-makers and policies a capacity for regulating and for "regulative" programs. Low-cost, simple pollution abatement technologies allow government to regulate an industry without encountering severe political opposition from those firms. Contraceptives are one foundation of an effective population policy.

However, the analysis of distributive, redistributive, self-regulative, entrepreneurial, regulative, and communal secur-

Science, Scientists, and National Policy Making

ity arenas implies that *all scientists' contributions will not be equally useful to policy-makers.* Policies with impacts that politically significant groups' perceive to be "distributive" or "communal security" oriented might be more attractive. *Scientists' contributions to distributive policies should then be used and stressed by policy-makers. When scientists' contributions themselves are distributive and do not require changes in people's life styles, patterns of behavior, or resources, they too are more useful to policy making.* The distribution of benefits, communal security, and values is politically productive. But redistribution and regulation harbor much political risk and peril.

Still, if a policy-maker must set forth policies that will be received as "redistributive" or "regulative" by politically significant groups, he ought not to shy from using scientists in all the capacities he can. They may augment his actions with justification, arguments, facts, and technologies for action. *Self-regulation can be threatened, regulation proposed, and redistribution suggested when the policy-maker is fully armed with scientific evidence, effective technologies, and persuasive estimates of the benefits resulting from regulation, redistribution, or an end to self-regulation.* Scientists can help to reduce the political risk and increase the chances of success.

Scientists can *assist the government to improve its effectiveness as an institution.* Members of a governmental bureaucracy are naturally reluctant to sanction reorganizations or new policy-making techniques. Such changes might alter the existing distribution of power and respect in the organization. However, the interests of effectiveness and long-run survival of an organization, even government, demand continual changes in its internal procedures. Reorganization and new techniques thus become "investments" in future usefulness and effectiveness. Even though techniques like PPBS, information systems, policy sciences, game theory, and cost-effectiveness may strengthen the positions of top managers and scientists who use and are served by the techniques, they may also make the whole organization more viable.

Scientists therefore can organize, staff, and operate research units in government; implement new techniques of decision-making; improve the internal organization of bureaus and agencies; and provide studies of the impact and costs of existing and alternative programs. On balance, *such uses of scientists may alter the balance of power within the government, but the total institution will become more effective.* The hands of Presidents and Secretaries may be strengthened in the bureaucracy's internal politics, but even those whose power is diminished will find solace in a more secure future for government itself.

Implications for Science Policy-Makers and the Formation of Science Policy

Michael D. Reagan has described "policies for the development of scientific capabilities and resources, and policies for the utilization of those resources in problem-solving situations." [9] The manner in which we allocate funds to the sciences (policies for the development of scientific capabilities and resources) will have a fundamental impact on the "utilization of those resources in problem-solving situations" and policy making.

If we do not fund the social sciences, and then suddenly decide we need them, they will have neither the capacity nor capability to help policy-makers. Similarly, lagging funds for physicists and chemists would cause many months delay, perhaps even years, should the nation suddenly need new pesticides or antipollution technologies. Policy-makers probably can always find scientists to render advice and provide devices, but whether they will be "competent" and "effective" is another matter.

Congressmen and Presidents have recently stressed the use of social scientists and social science in solving social problems. Yet not only do some political figures fail to see the social sciences as "competent," but social scientists may actually lack the knowledge and "science" required for meeting Congressmen's and Presidents' expectations. Michael D. Reagan has foreseen this problem:

> The danger now, in fact, is that legislative enthusiasm may outrun the ability of the social scientists to perform the miracles being asked of them in such fields as race relations, slum rebuilding, and population control.[10]

A multitude of *criteria* have been proposed as justification for supporting science and particular scientific fields.[11] They include the following, as well as many others:
1. Create possible technological spinoffs.
2. Enhance the possibility for solving social problems, aiding the "disadvantaged," and cutting social costs.
3. Stimulate the economic growth rate.
4. Promote "science for its own sake" or science as a "creative" activity like the arts.
5. Fund impending scientific breakthroughs, which otherwise may not occur.
6. Strengthen the position and power of either the government or private sector.
7. Meet the demands of the marketplace.
8. Promote a nation's national security, economic position in world markets, and international scientific leadership.
9. Develop a "tertiary industry" of ideas, mental activity, and science to supplement primary production (raw materials) and secondary activities (manufacturing).
10. Enlighten, guide, and improve the effectiveness of public policy-makers and policy.

Variously using selected criteria, the United States has stressed funds for science related to war, space exploration, atomic energy development, and health. These promote the nation's communal security against foreign foes and domestic fears. Increasingly, policy-makers have laid emphasis on fields ripe for breakthroughs and relevant to policy formation. Thus support for the social sciences, molecular biology (genetics), technology assessment, oceanography, and policy studies have increased.

But participants in science policy making have only begun to consciously make choices. Policy for scientific research and development remains largely ad hoc, tempered to the demands of leading scientists shaping policy and fluctuating with the political mood of the nation. We have yet to systematically link support for scientific research with the use of the research in problem-solving situations.

No one really knows how certain science policies distort the scientific community and scientific research. There is the possibility that post-Sputnik emphasis on increased scientific manpower has diluted the quality of science and scientists. Perhaps the preference for large-scale research has unduly forced "team research" proposals submitted by less-innovative and creative scientists. What have we done to universities' autonomy? Has undergraduate teaching fallen in quality when scientists and better-qualified graduate students left the classroom for the research grant? Perhaps the largesse of funds for the "hard" sciences has maligned a balance of power within the scientific community, allowing the "rich" to assume policy-making positions that enable them to continue cutting off funds to the "poor." [12]

Have not these very demands for utilization and relevance to policy making damaged scientific progress? Scientists may frame research and proposals for funds in order to receive support. But should not science progress because scientists' objective is a greater store of knowledge? Must there always be a known use when the research begins? Should the scientific community have to "sell" the "benefits" of science to political leaders and the public? How much freedom should we allow the scientific community to pursue whatever scientific objectives it deems appropriate? Research can help leaders' answer these questions.

Political leaders, scholars, and scientists have bantered these questions and criteria about for many years. But now that we have probably raised all the questions, it is time to use these criteria systematically in forming science policy. We need not require a specific use in order to fund research, but there should be a way to specify and anticipate the benefits accruing from resources committed to science. And above all, we will need scientific competence and excellence. That will mean more money, but it will ensure that whatever our objective it will be well met.

Policy-makers will have more "competent" advice and technology. Aesthetically, creative spirits will more fully flower, and the sense of "knowing" will be sharpened. National security and economic markets will be served. And science itself will develop.

This requires a conscious, systematic effort to form policies for the "development of scientific capabilities and resources" and particularly policies, instruments, organizations, and mechanisms for using scientific research and resources in making public policy. Concurrently, policy-makers must develop an appreciation of the relationship between these two faces of the science and policy-making relationship.

Notes

1. Yehezkel Dror, *Public Policymaking Reexamined* (San Francisco: Chandler Publishing Co., 1968), pp. xi–10.

2. Robert H. Salisbury, "The Analysis of Public Policy," *Political Science and Public Policy*, ed. by Austin Ranney (Chicago: Markham Publishing Co., 1968), p. 168.

3. Salisbury, *op. cit.*

4. Murray Edelman, *The Symbolic Uses of Politics* (Urbana: University of Illinois Press, 1967), pp. 23–29, 36.

5. Anthony G. Oettinger and Sema Marks have shown how the failure of technological solutions to human problems has also affected education. See *Run, Computer, Run* (Cambridge: Harvard University Press, 1969). Computer teaching systems and learning machines are rigid and cannot meet individuals' needs, but the fascination with technological solutions goes on.

6. Theodore Lowi, "American Business, Public Policy, Case Studies, and Political Theory," *World Politics*, XVI : 4 (July, 1964), pp. 677–715. A concept associated with Lowi's "arena" is the "issue-area." See James N. Rosenau, "Foreign Policy as an Issue-Area," *Domestic Sources of Foreign Policy*, ed. by Rosenau (New York: Free Press, 1967), pp. 11–50, and Stephen J. Cimbala, "Foreign Policy as an Issue Area: A Roll Call Analysis," *American Political Science Review*, LXIII : 1 (March, 1969), pp. 148–156.

7. See Salisbury, Lowi, Rosenau—*op. cit.*—and also Theodore Lowi, "Making the World Safe for Democracy," *Domestic Sources of Foreign Policy*, ed. by James N. Rosenau (New York: Free Press, 1967); Robert Salisbury and John P. Heinz, "A Theory of Policy Analysis and Some Preliminary Applications," *Policy Analysis in Political Science*, ed. by Ira Sharkansky (Chicago: Markham Publishing Co., 1970), pp. 39–60. Robert Eyestone, "The Life Cycle of American Public Policies: Agriculture and Labor Policy

Since 1929" (Unpublished paper presented at the 1968 Annual Meeting of the Midwest Political Science Association, Chicago, Illinois); and Brian R. Fry and Richard F. Winters, "The Politics of Redistribution," *American Political Science Review*, LXIV : 1 (June, 1970).

8. See V. O. Key, *Southern Politics* (New York: Random House, 1949), pp. 298–311; Raymond Dawson and James A. Robinson, "Interparty Competition, Economic Variables and Welfare Policies in the American States," *Journal of Politics*, XXV : 2 (1963), pp. 265–298; Thomas R. Dye, *Politics, Economics and the Public* (Chicago: Rand McNally, 1966); and Richard Hofferbert, "The Relationships Between Public Policy and Some Structural and Environmental Variables," *American Political Science Review*, LX : 1 (March, 1966), pp. 73–82.

9. Michael D. Reagan, *Science and the Federal Patron* (New York: Oxford University Press, 1969), p. 152.

10 *Ibid.*, pp. 121–122.

11. An excellent and thorough discussion and criticism of various criteria is in Reagan, *op. cit.*, pp. 270–302, 34–70. Some original key sources in the debate over "criteria" are collected in Edward Shils (ed.), *Criteria for Scientific Development: Public Policy and National Goals* (Cambridge: M.I.T. Press, 1968). Shils has collected articles from *Minerva*.

Three important contributions are Alvin M. Weinberg, "Criteria for Scientific Choice," *Minerva*, I : 2 (Winter, 1963), pp. 159–171; Stephen Toulmin, "The Complexity of Scientific Choice: A Stocktaking," *Minerva*, II : 3 (Spring, 1964), pp. 343–59; and Toulmin, "The Complexity of Scientific Choice II: Culture, Overhead or Tertiary Industry," *Minerva*, IV : 2 (Winter, 1966), pp. 155–169.

12. Reagan, *op. cit.*, pp. 303–319.

PART TWO.
SCIENCE, SCIENTISTS, AND PUBLIC POLICY MAKING: THE 1970s AND THE YEAR 2,000

The Future of Scientists, Science, and Policy Making: Issues, Tensions, and Prospects

CHAPTER 8

Man could not do many things nor have many things without science. He turns to his scientists for tools and possessions contributing to the fulfillment of his needs. And often, because he is filled with insecurity and fear, he defers to his scientists for advice, direction, and solutions holding promise of security and hope.

But the scientist needs man as patron, customer, and subject matter. The interdependence of man and scientist is an established fact. This chapter deals with their relationship, assesses its future, and indicates problems that might plague scientists, societies, and public policy making. What does the experience of scientists' participating in postwar policy making (1945–1968) in Washington indicate for the future?

NOTES TO CHAPTER 8 START ON PAGE 291

What new areas or undeveloped regions are ripe or waiting for scientists' attention and participation? How will scientists contribute to policy making in less developed geographical areas, Congress, the courts and legal system, and the states and cities? What tensions and issues have science and technology begun to create for policy making and public policy? In what respect can scientists both create and solve problems? What about pollution, genetic changes, population growth, deprivations of minorities, smoking, drugs, and sprawling cities? Can we assess and control the impacts of science and technology? What threats do society and government pose for science and scientists? How can men use but control, shape, and channel scientists' contributions and participation? What tensions arise for a society with both scientific and democratic aspirations? Can scientists parley knowledge and skill into power, wealth, and influence?

Such questions demand answers. Probably I can suggest some here. But primarily the questions must be asked, issues identified, and future outlined. Society and the public policy-making system must provide the answers. And good answers depend on appropriate questions based on accurate statements of problems.

THE UNDEVELOPED REGIONS:
SCIENCE, SCIENTISTS, AND POLICY MAKING

While scientists' involvement with policy making in the Executive Branch has been notable, several areas of possible participation have lagged. These include the Congress, states, cities, developing areas, and courts and legal systems. And there are reasons for scientists' absence in these areas, factors that may or may not disappear in the 1970s. What about those regions that might be fruitful areas of scientists' involvement? What does their postwar experience in Washington (Chapters 1–7) say about scientists' future in these fields?

Congress

Scientific advice for the United States Senate and House of Representatives has been quite random, ad hoc, and poorly supported. But scientists and scientific advice have not been absent from the halls of Congress, and the situation is getting better.

Many Congressmen have little understanding of science, scientists, or the contributions science can make to a policy-maker. Their understanding of the social sciences continues to be a major difficulty, since many legislators do not see them as "scientific" and feel that any man can be his own "social scientist."

Congressional staffs are undermanned, poorly paid, and operate on low budgets. They have little money for financing full-scale research in universities or private research organizations. As a result, Congress depends heavily on scientists attached to the Executive Branch for testimony and advice. Thus, even in the field of scientific advice, Congress faces a dependence on the Executive Branch for information, intelligence, and advice. And the Executive Branch prefers to retain its own prerogatives and advantages in this situation. Therefore, administrators shy away from common efforts with the Congress to coordinate or integrate their communications and policy-making apparatus for scientific matters.

Traditionally, Congress's relationships with science and scientists have involved hearings and investigations related to national security, loyalty, and scientists; legislative oversight of agencies funding scientific research and making science policy; and ad hoc use of scientists for testimony at hearings. Recent developments have included legislators' concern with the geographical distribution of science and research funds (cf. location of the particle accelerator at Weston, Illinois) insofar as they have become pork-barrel matters like any project for representatives and senators' constituencies. Some social scientists have been involved in the area of Congressional reform and reorganization.

260 The Future of Scientists, Science, and Policy Making

Oversight has increased in recent legislative sessions. Congress has trimmed the overall science and research funding budget, eliminated specific programs (cf. cancellation of the Mohole drilling project), stressed spillover or transfer of research benefits to the civilian sector, required some social relevance or policy contribution from research, and expressed concern over "inequities" in the distribution of research funds (cf. the endowment of California, New York, and Massachusetts).

But Congressmen face serious problems in using scientists' advice. They can make political decisions on science policy and research funding, but their capacity for absorbing and digesting scientists' contributions to public policy debates remains limited. Congressmen, like citizens and scientists outside a specific scientific field, have difficulty comprehending scientists' arguments. And when scientists' testimony or recommendations conflict, the dilemma is even more humiliating. The befuddlement of many Congressmen during the 1969 ABM Safeguard debate illustrated this situation. Increased staff advice and background information can be helpful, but the technical complexity of scientific fields will probably always outrun nonscientists' efforts at comprehension.

Congress has taken some steps to augment its capacity for processing and using scientific information. Traditionally, science affairs were handled in the Senate Commerce Committee (Weather Bureau, National Bureau of Standards), House Merchant Marine and Fisheries Committee (Coast and Geodetic Survey), and the Joint Atomic Energy Committee (Atomic Energy Commission). But since 1946, the number of executive agencies involved with science has increased more than a hundredfold. Almost every Congressional committee now must consider scientific matters.

New committees have arisen to fill the vacuum. They include the Senate's Committee on Astronautical and Space Sciences, the Senate's Government Operations Subcommittee on Government Research (Sen. Fred Harris, D., Okla., Chm.), and the House's Committee on Science and Astronautics with new responsibility for NSF, NASA, and the National Bureau of Standards. The Committee on Science and Astronautics has three recognized subcommittees in the area—Science, Research and Development (Rep. Emilio Q. Daddario, D., Conn., Chm.), Research and Technical Programs (Rep. Henry Reuss, D., Wisc., Chm.), and Intergovernmental Relations. Meanwhile, the Joint Economic Committee, created in 1946, continues to use economists' advice.

Congress created a Science Policy Research Division in the Library of Congress' Legislative Reference Service in 1962. Its productivity has already been notable. There have been proposals for a Joint Committee on Science and Technology—a joint staff that would coordinate the ad hoc use of scientific experts—and more funds for science-related committee staffs.

The Congress has never been organizationally equipped or amenable to outside information or expert advice. The problem with scientists thus is not new. But because science and scientists have become so infused in the governing and policy-making process, Congress has no choice but to increase its capability for using their contributions. And, inevitably, we can expect it to do so. Congress' survival as a viable, politically meaningful institution hangs in the balance.

Generally, law making is a "regulative" matter. Scientists' relationships with Congress have traditionally revolved around "distributive" situations (research funding, project locations) or "symbolic" legislative activities (security-loyalty scandals and hearings). Now, Congress has begun to absorb some "regulative" and "redistributive" concerns previously confined to the Executive Branch. As Congress moves into policies with these impacts, scientists will become more important. And as it becomes more concerned with communal security (health, weapons, defense), management of the economy, and regulation (pollution, safety, conservation), its use of scientists must surely increase. An aggressive, active Congress will need and use more information, intelligence, and expertise. Concurrently, scientists' influence on Capitol Hill can be expected to grow.

The States and the Cities

Scientists traditionally have been rather uninvolved with American cities and their role in the "distributive" politics of state policy making has been only somewhat more noticeable.

Frederic Cleaveland's study of support for scientific research in six states *(Science and State Government)* shows the states' use of scientists and research.[1] Basically, research funds have been provided in four areas of state concern—agriculture; resource development and public works; higher education and research; and health, education, and welfare. Cleaveland found that social, economic, and political factors shape states' division of funds. New York tended to give more support for research in the area of health, education, and welfare. New Mexico supported applied research more heavily. California, New York, North Carolina, and Connecticut stressed basic research and higher education. Wisconsin and New York tended to give more funds to research relating to agriculture, and New Mexico emphasized resource development and public works.

The various political and economic situations in the six states apparently produced these differential levels of support for research. In fact, rural dominance of most state politics and policy making leads to a general agreement that agriculture-related research will be significantly supported. Similar reasoning explains states' encouragement of scientists involved in fields such as public health, mine safety, and resource development (wildlife management, geology, and forestry). Recently, many states' efforts for industrial development have doubtless led to increases in state funds for applied research and development facilities. Increased competition for federal grants and contracts have had identical effects.

But states have lagged in their use of science and scientists. Though they may fund scientific research, states have failed to use scientists and science in forming public policy. A notion of scientists' contribution to policy making has been lacking. Cleaveland notes that the states "underemphasize this policy-supporting dimension." [2]

Reasons behind the undeveloped scientist and policy-making relationship in states and cities are manifold:

1. The national government's use of scientists has been stimulated by military needs and insecurity. States have had little such obligations and responsibilities.
2. States, as distinguished from Washington, have preferred general political administrators rather than technical specialists for major policy-making responsibility.
3. Washington and the federal government have more prestige and are more attractive to the scientist wishing to be involved with policy making. Scientists prefer consultation in Washington over consultation in Harrisburg.
4. The federal government first invited scientists to participate in policy making. Scientists wishing to be active on a state level had no such invitation and had to gain access through the media, lobbying, and obviously political activities.
5. Support for scientific research has flowed from Washington, not from state capitals and cities.
6. State science advisory groups are politically exposed and, as protective devices, have been saddled with an incapacitating balance of geographical and major institutional representation (regions, industries, agriculture, types of universities, etc.).
7. State science advisory groups have been located down within the bureaucracy and not given top-level status under the governor.
8. Lack of adequate staff and funds has limited the activities of state science-related activities and groups. States, caught in a financial squeeze, can be expected

to cut research and development expenses just as firms cut back capital investments.

Harvey M. Sapolsky, who identified and thoroughly discussed these several factors, has suggested that the situation is improving.[3]

More and more states are creating mechanisms for formally locating scientific advice within policy-making circles. And states and cities have more responsibilities placed upon their shoulders than ever before. The traditional activities in agriculture, disease prevention, resource exploitation, wildlife and forest management, and safety have grown, and more burdens have been added. The threats of pollution and urban sprawl and the opportunities of atomic energy, sea resources, and transportation have become matters for states' policy responses.

States have begun to realize that they need the capability that scientific advice and expertise can provide. They need the capacity to act, and to act appropriately, in solving state and urban problems. Thus New York City hired the Rand Corporation for a systems analysis of some administrative aspects of its government. States may begin to use systems engineers on the dilemmas of urban, ghetto, and transportation problems. Granting agencies, science advisors, scientific committees (cf. PSAC), and formal consultantships emerge within the confines of state and urban governments.

States and cities are moving beyond the traditional "distributive" functions of providing funds for agricultural research, disease prevention, resource development, and education. No longer can highways and mass transportation systems be planned solely on the basis of *political* considerations. No longer can industries be allowed to pollute lakes, streams, and the atmosphere.

Increasingly, states and cities are asked by citizens and politically significant groups to regulate and redistribute as well as distribute benefits. Policies with distributive impacts are not difficult or politically perilous to form. Scientists would not be needed in policy making. But when a state or urban government cannot afford to be wrong, needs a realistic diagnosis of a problem, requires a technological solution, or needs to strengthen and justify its case and position, scientists become most useful. The urban crisis, demands for regulation of pollution, and the perils of regulatory actions indicate that states' and cities' use of scientists will increase. And scientists' influence might reasonably be expected to grow in these previously undeveloped regions. The lag between national and state or urban government will continue, however. Washington, not Harrisburg, has the responsibility for national defense and military security, which stimulates so much of a government's use of scientists.

The Courts and Legal System

Judges traditionally rely on legal precedents, reason, and tradition as a basis for their decisions and popular support of those decisions. Science and scientists' contributions are minimized in this type of legal system.

But courts have often needed and accepted scientific evidence that added to their knowledge of the situation under adjudication. Criminology has a long history in the courtroom. Economic evidence on firms' shares of markets contributed to antitrust cases. Data on wages and hours was involved in labor and industrial litigation. Psychiatric testimony has been accepted and used in certain cases. Statisticians and political scientists' work on malapportionment was involved in the series of reapportionment decisions. Evidence on genetic and chromosomal make-up of defendants accused of sex crimes has even been introduced in trials. The physical sciences have contributed in patent litigation and similar issues.

Sometimes, scientific evidence has no impact on judges' decisions. Scientific arguments may only be inserted in the opinion to justify or further buttress an argument or decision based on legal precedent, interpretation, or reason. Thus sociological and psychological science cited in *Brown v. Board of Education* in 1954 may have been more oriented toward strengthening public acceptance of the decision than convincing the justices themselves. Still, it would be difficult to tell to what extent these scientists' influence stemmed from the climate of tolerance and "equality" they helped to create.

Detailed study would have to be made, but initially there seems to be some basis for viewing the courts as a province of the behavioral sciences. Their role there appears to be as predominant as the physical scientists' activities in the legislative and executive branches perhaps because so much civil and criminal law involves *individuals'* freedom, well-being, life styles, and personal relationships, and because courts stress evidence as fundamental to decision making.

Courts do not hire scientists. They do not contract out research. This responsibility belongs to litigants and their lawyers. Thus scientists' influence on litigation comes indirectly through lawyers' arguments or scientists' testimony.

Richard C. Cortner has argued that litigants challenging the legal status quo or precedent tend to use nonlegal, outside materials in their arguments. He cites the NAACP's use of psychological and sociological data regarding the effects of segregated education in *Brown v. Board of Education*, and business interests' reliance on the social and economic theories of Adam Smith and Herbert Spencer in their efforts to convert the Fourteenth Amendment's due process clause into a substantive limitation on the exercise of state power.[4]

Cortner's "aggressive" litigant has no precedent on his side and, as a challenger, must compensate for that deficiency. This difficulty usually accompanies the cases of "disadvantaged" litigants, whose rights have never been meaningfully represented and tested in the legal system. And not infrequently, when earlier cases establishing precedents were decided, knowledge and scientific evidence may have contradicted existing evidence (theories of racial inferiority). Or scientific evidence might not have been known or available.

The legal system and its courts are invariably involved in "regulative" or "redistributive" situations. Courts imprison people, require people to pay money to other people, divide values, and assign basic rights to previously deprived groups. Such decisions are likely to arouse the interests of concerned parties, some of whom will not benefit from the courts' decisions. Science and scientists' contributions can help because they give the court additional justification for its decision, supplementing the traditional justifications of legal reasoning, autonomy of the court, and precedent.

Periods of rapid social and technological change will increase pressures on the legal system. More cases will arise and more fundamental assumptions challenged. Indeed, an economically progressing system can afford more litigation, and more will be demanded because more and more groups have access to resources for court action. Such opportunities for disadvantaged groups were not previously available.

Furthermore, science quite often will intervene or contribute in the interest of a disadvantaged litigant. This may make scientists less popular. Consider the public's attitude toward introduction of scientific evidence in criminal cases. Scientists have increased society's sophistication in treatment of crimes. Geneticists suggest that criminal tendencies may stem from parents' genes. Psychiatrists suggest that murder may result from insanity rather than willful, rational action. Criminals freed on such reasoning, after all, threaten citizens' sense of communal security, law, and order, and weaken the "forces of peace." Security has been compromised for a "redistribution" of equality. Citizens are less secure, while criminals receive a fair shake. And the courts and scientists are embroiled in increasing controversy.

What about scientific studies of capital punishment as a deterrent to capital crimes? What about "hanging juries" and juries limited to certain races? What should a court do about such evidence? How can a judge evaluate the "scientific" merits of the evidence? Whose side does the evidence favor? Questions such as these will become more and more pertinent, because the complex and unique relationship between scientist and legal system will certainly become more developed in the 1970s. How will courts balance legal and scientific factors? Will cases be decided in a court of science rather than a court of law? What evidence will be acceptable and how will it be used?

Developing Regions

Scientists' skills and contributions have not been applied in less-developed areas of the United States.[5] The state of native scientific research in Appalachian and some Southern areas has become a matter of national concern. Development of the quality and quantity of science in state-supported universities in some regions has attracted the attention of Congress and the Executive Branch.

Many such regions lack scientific traditions, skilled manpower, adequate high school and undergraduate educational systems, resident scientists, desire for scientific development, and a public appreciation for the benefits of science. No concentrations of industries or economic enterprises make demands for applied research facilities in the universities, state governments, or private firms.

And yet these pockets of underdevelopment have great potential, which science might free for growth. They have labor, natural resources, fuels, and room for expansion. But manpower must be educated and trained, water and fuel resources opened up, people fed, new industries capitalized and nurtured, and a whole economic infrastructure constructed. Science and technology can help make this transformation possible. But these scientists and facilities must be located within developing regions.

Scientists in such situations become instruments assisting government in "distributing" economic benefits and promoting development. But the development process, when structured and guided by government, must have adequate scientific advice and direction. Otherwise, the effort may fail, shipwrecked on the rubble of short-run, political choices. Scientists' participation in this development process will be not unlike their involvement with Washington from the 1800s through World War II, limited to the essentials of nation building and the "distribution" of economic progress and communal security.

EMERGING ISSUES AND PUBLIC POLICY MAKING: THE SCIENTISTS' CONTRIBUTIONS

Scientists' behavior as scientists, political participants, and citizens has led to a multitude of problems now confronting, and soon to confront, society and public policymaking. But scientists' are not alone in accepting responsibility for these problems resulting from scientific and technological innovation. Many others share the burden. Scientists may only have made man's destructive and trouble-creating choices possible.

Scientists, many as unwilling silent partners, and societies have used science and technologies emerging from science. The results have not always been acceptable to later generations. They have not always been acceptable to men who made the decisions. And many are the "difficulties" that piggybacked along with beneficial outcomes.

Rapid, uncontrolled industrialization, moving on a rising supply of science and technology, produced a polluted environment. Was this a price we would have paid for the benefits of rapid industrialization? Could we have anticipated the pollution crisis and had "our cake and eaten it too"? What about agricultural productivity and pesticides, such as DDT, that made an abundance of "polluted" food possible?

Medical science made it possible to control disease, treat illness, and improve sanitation. But men lived longer and contributed to the population explosion. Scientists and World War II ushered in nuclear discoveries and technologies. Doctors could therefore treat cancers with radiation, but society also reaped the difficulties accompanying nuclear weapons and radiological pollution.

Computers and television allowed broadcasters and political experts to predict election results minutes after polls closed. Thus voters in California would know Connecticut's choice for President hours before their own polls closed. And Southern Californians would know returns from Northern California before their own polls closed.

Science helped society become so affluent that its waste products (sewage, trash, and junk) became an acute problem. Technologies of inspection and surveillance have led to invasions of privacy, penetrating heretofore insulated sectors of human activity. Government itself has used these techniques to inject its nose into its citizens' private affairs. Computers have made data centers possible and created the "threat" of central files on an individual's medical background, personal characteristics, credit rating, financial affairs, and voting behavior.

New technologies have created labor and management problems and forced changes in traditional collective bargaining arrangements. Typographers on newspapers have faced new typesetting technologies. Prefabricated housing has threatened building trades unions. Cities' industry and union-dominated building codes have become obsolete in the face of new construction materials and technologies. Such codes have inhibited the construction of new low-cost urban housing facilities.

The technology of fluoridating water raises bitter community conflicts. So may the irradiation of foods or the use of pesticides trouble public policy making in the 1970s. Drugs, either legally dispensed by pharmacists with government approval or injected into the body in a dark alley, have become a pressing matter for public policy. Heroin and thalidomide could both lead to disaster, one clandestinely and the other with government sanction, non-interference, or failure to act.

Technologies of reproduction and communication have

Emerging Issues and Public Policy Making

not only enriched the Xerox and 3-M Corporations. They have mandated changes in the nation's patent policy and copyright laws. When community antenna television systems (CATV) borrow programs from the air and relay them to customers, what are their obligations to producers and the originating station? What challenge do they pose to copyright laws and FCC rules on communications? What is the effect of United States' patent policy on innovation, discovery, and creativity? Who owns the patent on a discovery made by a researcher paid with government funds?

Scientists themselves have caused problems for policymakers. May a behavioral scientist secretly, with the judge's permission, bug a jury room to study the behavior of a jury? Should the American government allow its State or Defense Departments to support scholars' research conducted on foreign soil into the behavior and internal politics of foreign countries? What are the costs of fiascos such as Project Camelot, where an inquiry into ways of avoiding or combating a left-wing revolution in Chile and other countries led to a full-blown scandal in Chile and Washington? What are the "rights" of individuals and groups subject to research? What about prisoners who volunteer to participate in drug tests or to be given diseases? To what extent is the ten-year Census an invasion of privacy? What questions can the Census justifiably ask? How can management or political executives justify hiring organizational scientists or management research firms to conduct studies aimed at increasing workers' productivity and the firm's profit? Does a man have a right *not* to be studied, particularly when the research may lead to his death, misuse, or manipulation?

Science has improved our capacity to "change" individuals' sex. What will be the response of public policy? What about laws on homosexuality? Medical science may soon perfect capabilities for creating babies in test tubes, selectively breeding children from particular sperm and eggs, and controlling the sex of offspring. What problems does this capability, perhaps used in conjunction with sterilization and contraceptive devices, pose for a nation's population policy? Does *every* man have a right to mate in order to produce offspring? Who will decide? How?

But there is "good" and "bad" residing in every scientific or technological possibility. Nuclear science has led to bombs and radiation therapy. Agricultural science allows farmers to increase their yield and pollute streams with pesticide runoff. Behavioral science produces greater understanding of racial and cultural minorities and simultaneously produces knowledge that can be used to manipulate such groups. Medical science extends our life span, creating more population pressure, and develops cures for dreaded diseases.

Perhaps the problem is one of predicting and anticipating the future, as many now suggest. Rep. Emilio Q. Daddario (D., Conn.) has proposed a Technology Assessment Board, which would concentrate on foreseeing the consequences of technological and scientific change. Both beneficial and detrimental effects would be located and, hopefully, measures taken to alleviate or eliminate the unwanted results.

Most societies are not willing to forego progress in order to avoid its pains and harmful results. But they can make efforts to anticipate problems, and, rather than progress in an uncontrolled fashion, plan for difficulties with an eye toward making sure they do not develop. Technology assessment would provide an "early warning system" that might help societies structure their futures. We might have begun research on methods of population control much earlier had we anticipated the results of rising life expectancies. Now society and public policy-makers must react to a problem already well established and out of control. Pollution might also have been anticipated. What about the problems of jet-age travel, supersonic transports, use of pesticides, housing technologies, computers, television, heart transplants, new definitions of "death," genetic control, and food supplies?

Technology assessment and the anticipation of problems is important because science gives society the capacity to

Emerging Issues and Public Policy Making

solve many problems stemming from scientific progress itself. Science and technology have led to air and water pollution, and pollution stemming from noise, radioactivity, waste, and thermal effects. But simultaneously, science is capable of developing means for controlling and abating pollution. In this sense, scientists can be both creators and solvers of social problems. And as such, they can contribute accurate diagnoses of public problems and possible alternative solutions for public policy-making.

Desalination of sea water may solve the problems of arid lands. Controlled-fusion reactions may enable production of energy from heavy water, now plentiful in the oceans. Drugs aid medical science in treating alcoholism, uncontrolled smoking, and mental illness. Medical sex "changes" represent techniques for treating personality abnormalities. Methods of controlling the sex of unborn children may relieve population pressures, since families would not have to "try again" for a male heir or daughter. Changes in an individual's genetic code or make-up may alter his personality, improve his memory and learning capacity, or alter his behavior. A society might medically "reform" its criminals and control deviant behavior. But who will decide what is "normal"? What will "Superman" look like? How will he behave and see the world?

Obviously, possibilities do not come without pitfalls and potential problems. I doubt that man will soon develop a capacity for saying "no" to possibilities. But at least we can expect that he will anticipate unwanted effects, seriously weigh fundamental political and moral questions that will be raised, and forge policies based on a full consideration of the "pluses" and "minuses" of all alternatives.

A greater tragedy would be to stand aside, uttering "que sera sera," and doing nothing. Full consciousness during man's descent into Hell may not be comfortable, but it is more morally defensible than a blind ride as if on a playground slide.

Society must use its scientists, for they can contribute

to policies designed to cope with the society's problems. Such use of scientists may mean the difference between a society's success and failure. Still, some sciences and technologies can offer more to public policies than others.

Technologies perceived by politically significant groups as "distributive" will be more politically attractive and feasible than solutions seen as "redistributive" or "regulative." And some technologies are indeed "redistributive" insofar as powerful groups react to the technologies' impacts.

Physical technologies, or operating systems comprised of material, nonbehavioral components, tend to be seen as "distributive." At most, they merely require individuals or groups to use them passively or allow their use. Examples already considered as components of public policies include street lights as a crime deterrent, weapons systems, contraceptive devices, smog control devices, seat belts, and desalination techniques.

But *behavioral technologies,* or types of human relationships and patterns of behavior, are more likely to be seen as "regulative" or "redistributive." They may require changes in life style, prohibition of some activities, involvement, or threats to important values. Examples include techniques of decision and policy making, self-analysis, reapportionment, antitrust actions, dieting, civil defense shelter programs, safe-driving campaigns, the draft, and the "rhythm" method.

Fostering the development of *physical technologies* is generally politically advantageous in terms of support and avoidance of conflict. Such technologies permit societies to attain goals and realize material objectives. Government, desiring to regulate a firm's pollution of air and water, may offer a new technology of pollution control sufficiently inexpensive and simple that compliance with the regulation is easier. These types of technologies, properly designed, enable the regulated group to adjust to the deprivation and perhaps view the government's action as "distributive" because of the physical technology it made available. Tranquilizers, antismoking pills, dieting pills, and the proposed

Emerging Issues and Public Policy Making 275

air bag for automobiles enable individuals to achieve ends without changing their life styles or patterns of behavior.

People desiring to lose weight will accept a solution in the form of a pill but are much less likely to alter their eating patterns or "starve" themselves. Similarly, tranquilizers or personality-control drugs avoid the necessity of intensive self-analysis, the agony of self-awareness, or a slower pace of living. Contraceptives likewise permit couples to achieve or maintain a preferred pattern of sexual behavior, without arbitrary self-regulation or discipline. In this sense, technologies free men from regulation or alteration of preferred ways of living.

Physical technologies are also a means to solving social problems without a commitment of time or personality. They are impersonal. Men will spend money for emission control devices on automobiles, but they are less likely to switch to slower and smaller electric cars. Air strikes on enemy targets are preferable to assaults on foot, because the air strike is quick, impersonal, and involves no substantial commitment of human life or personality. Indeed, Americans seem to prefer expensive weaponry to the draft and the foot soldier's very personal involvement in war. Physical technologies are "sanitary" means to desirable ends, when many means are so personal and require so great a commitment of time and personality as to be prohibitive. Money can buy physical technology, and as distasteful as taxes and donations may be they are still preferable to giving of one's time and committing one's personality or life.

But much *behavioral technology* evokes the specter of regulation or redistribution. Any new technology, physical or behavioral, implies change, and this is even implied in "distributive" situations. But change that *appears* to be redistributive and regulative is less politically attractive than "distributive" change. And change achieved without threatening human values or styles of living poses less political risk to the policy-maker.[6]

All of this *does not mean* that we should not use some

scientists and technologies, while neglecting and disregarding others. It *does mean* that we need to give attention to the varied contributions of scientists to public policies. Perhaps scientists themselves might frame their recommendations and research to achieve maximum usefulness for public policy making.

SCIENTISTS, SOCIETY, AND THE POLITICAL SYSTEM: TENSIONS AND DISTORTIONS OF SYMBIOSIS

Observers and sociologists of science have suggested many characteristics of societies that may facilitate and encourage scientists' pursuit of their *métier*. Indeed, these students of science and society have long debated the relative importance of these various factors.

Such characteristics or conditions [7] include factors stemming from society and the scientific community:
1. Economic surplus and the development of specialization, agricultural surplus.
2. Demands from capitalism or the industrial system for applied sciences and technologies.
3. Expanding commerce and trade.
4. Needs for a navy or sea transportation.
5. Fulfillment and/or demands for improvement of basic human needs, like shelter and food.
6. Military threats from other nations, needs of a military (land, sea, or air and space).
7. Stability.
8. Freedom from hierarchy, totalitarian closed systems, excessive political interference.
9. Freedom from being left alone (laissez-faire).
10. Welfare state, predominance of public and popular material goods and benefits.
11. Plural society, open communications network.

12. Freedom from patronage systems, civil service or seniority systems, class or caste-based systems.
13. Emphasis on "puritan ethic," savings, investment, stress on "this world" rather than "next world".
14. Dogma-free environment.
15. Secularization, stress on secular values and actions.
16. Shift from teleological and religious explanations to "cause and effect" or scientific explanations of cosmos.
17. Pragmatism.
18. Individualism.
19. Skepticism.
20. Information and communication explosion (books, printing press, journals, literacy, learned societies, abstracts, computer storage and retrieval techniques).

But these characteristics and conditions facilitate the growth of science. What events, conditions, and pressures threaten science's continued growth and productivity? How can society or social institutions distort, destroy, or delay scientists' development? Science is a social institution. By what means can one social institution affect another, particularly one so life-giving as science?

Society's harmful or distorting impact on science can limit and malign the contributions science can make to society, policy making, and public policies. The twin perils of mass society and closed totalitarian systems can threaten science and scientific development. Massocracy, or the rule of the masses, brings with it appeals to emotion rather than reason, unplanned spontaneous action, reliance on impulsive behavior rather than calm thought, and a sense of commonality that isolates and buries individual achievement in the group. The calm, deliberate, rational, and individually creative progress of science suffers in such an atmosphere. And the impulses of the mass may tug scientists' efforts in unproductive directions.

Totalitarian systems monopolizing the lives of citizens are equally threatening. Political considerations predominate under centralized rule through a "leader" or elite that uses ideology to manipulate and control individuals' behavior. Under such conditions, some segments of science may flourish. Weapons-scientists and specialists in the manipulation and control of individual behavior may receive vast funds. Yet it is still questionable whether the atmosphere of secrecy and arbitrary political decision making will allow even these "priority" fields to mature and produce as much as they are capable.

Leaders' demands that Soviet biology support the fundamental tenets of Marxism-Leninism led to the enshrinement of Lysenko, neglect of Mendelian genetics, and decades of stagnation in the Russian biological sciences. If Marxist man could be changed and transformed in a few years of his life and pass these newly acquired Marxist characteristics along to his offspring, then grains could be subjected to extreme heat and dryness and, when planted, grow in arid lands. Mendelian genetics had argued that mankind could not be so quickly changed, just as seeds could not be so easily altered. The Soviet Union had similar difficulties accepting Einstein's concept of "relativity," in part because it subtlely cast doubt on the Marxian absolute.[8]

Democracy itself harbors difficulties for science and scientists. Its faith in majority rule, distrust of planning, trust of the common man, and pressures for "equality" can produce an environment which sometimes is antagonistic to scientists. And politics in the open atmosphere of democracies can wreak its own brand of havoc. How much distortion and delay in scientific progress resulted from the United States' political commitment to land Apollo on the lunar surface by the end of the 1960s? What might science have done or contributed to policy making had that overwhelming objective not been dangled before citizens and scientists?

Many states have long inhibited or prevented agricultural research into butter substitutes, such as oleomargarine. This stemmed from the political pressures from farmers in these open, democratic, voter-bound systems. What, too, about America's commitment to build an SST or supersonic transport airplane? What about the vast R&D funds allocated to the so-called "military-industrial complex"?

Pressures come from the internal politics and traditions of government bureaus as well as from the open politics of elections, interest groups, and public opinion. Joseph W. Eaton has shown how "in-house" scientists closely attached to bureaus tend to do research that supports existing programs, avoids negative or discouraging results, and only *symbolically* avows the value of evaluative research.[9] Outside research, on the other hand, is more likely to evaluate the bureau *substantively,* question its errors, and criticize existing activities. Thus bureau scientists are more likely to share and support bureau norms, either because of pressure, sheer agreement, or a realization that their futures are dependent on "not rocking the boat."

Tradition can likewise cause difficulties. Ashley Schiff's incisive account of the Forest Service's rejection of scientific evidence on controlled burning and water run-off illustrates the impediments of tradition. The long dominant ideologies of "no-burning" and "no-cutting" had become so embedded in Service rules and thinking that new possibilities and alternatives never entered policy making and policy for decades.[10]

Religion and theology, posing alternative methods of arriving at truth and belief, have often impeded science's development. But their challenge has not just been limited to burning heliocentric astronomers at the stake. The famous Scopes trial and resistance to the Darwinian Theory of Evolution, continuing even into the 1960s, provide a more modern, though less deadly, example. Galileo's problems, as Bertolt Brecht's *Galileo* indicates, are not buried totally in obscure historical eras.

The emergent "garrison" and *security states* threaten their own brand of malignment of science and its contributions. We have "nationalized" and "socialized" science in a "public endowment of scientific expansion"[11] in order to increase our collective sense of military, economic, and personal *security.* How much of science's new-found largesse can be attributed to the "cold war" and Sputnik? What have we done to science's development in the process? How have we distorted science's future?

The *security state* co-opts scientists into the elite and integrates them into infinite numbers of "complexes" or amalgams. Scientists become members of the military-industrial complex, the "new industrial state," or the educational state. They join and link their futures to the elite and survival of the system. Their absorption is an unconscious process, but gradually society and its politics draw scientists into service of their needs. Whatever science might have done were it autonomous and free is lost. Scientists, like soldiers, are often mobilized for a cause not originally their own.

The development of science can be impeded and distorted by mass society, total systems, politics, and politically significant groups, religion, and elite-centered complexes or amalgams. But whatever the source of the distortion, scientists' contributions to a rational public policy are severely harmed. Society can use science and reap science's benefits only if society leaves science sufficiently free to pursue some of its own objectives.

Too many political demands and too much political interference may lead to poor science. And incompetent scientists and "bad" science cannot offer much to public policy making. Thus *bad politics* can be a source of *bad science*, constituting a major fault in the symbiotic relationship of scientist and socio-political system.

Scientists, Society, and the Political System: Mankind's Use of Science and Scientists

Public policy making must weigh and select among political, moral, scientific, ideological, cultural, and rational considerations. Thus the skills of lawyers, administrators, politicians, and other experts, as well as scientists, will continue to be essential in the formation of public policy. And, as in the case of lawyers and political specialists:

> The problem of scientific representation, then, is not a case of an omnipotent, ruling clique, but a significant case in the general problem of the relation of the expert in the policy-shaping process.[12]

The integration of experts, including scientists, into policy making demands rather careful consideration of means for shaping, controlling, and channeling their participation. Democracies cannot blindly defer to scientists' proof, since, as James Ward Smith argues, "the foundation of democratic political philosophy *consists in* the rejection of the concept of justification by proof."[13] Nor can we separate politics from science, confine scientists and their discoveries to the laboratories, and disregard science in our collective life.

Somehow, citizens and policy-makers must have the opportunity to hear scientists' case and accept "justification by proof." Democracies may continue to find justification for policies in majority votes, but each individual voter and leader must be free himself to accept and act upon scientific proof if he chooses.

But what about specific arrangements that provide control and direction for scientists' participation in policy making? We *cannot rely on traditional legislative and partisan politics to check and balance scientists' influence.* Neither the political parties nor legislatures are sufficiently staffed or knowledgable for such demands. *Voters or consenting citizens cannot adequately guide scientists' participation.* Scientists and their work cannot be subject to electoral approval. And the general public may often itself be unaware of problems stemming from scientific change. Pollution is an example. Or the public may vaguely see a need for accepting scientists' contributions but have no desire for changing its life style or paying the price. Cigarette smoking and automobile exhaust controls are examples.

The public, to some extent, can indicate after the fact whether "the shoe fits." However, so much scientifically based change (weapons, pollution) is either irreversible or requires great efforts to reverse. And science can be infinitely complex and hopelessly confusing to most citizens.

Nor can society or policy-makers defer to the scientific community for political and scientific advice. The scientific community itself is riddled with "politics." Though it has some appearance of a "hierarchy," blame would be difficult to fix on particular scientists. Responsibility cannot be easily located when research is done by teams, when science builds bit by bit on previous scientists' work, and when the scientific community is an amorphous group. Certain scientific fields are over-represented within the scientific community and hold disproportionate shares of influence within the group. Besides, scientists' behavior within their own scientific circles can be quite "political" inasmuch as many scientists pursue power over fellow scientists.[14]

Furthermore, what does a policy-maker do when a group of scientists gives him poor or malicious advice? They are not subject to reelection. They can be fired, but like lawyers returning to private practice, scientists retreat easily into universities. There is little promise in deferring to a scientific community that could be held accountable for its decisions.

Society cannot rely on competition and checking among its scientific, political, economic, cultural, and religious elites to shape and control its scientists' participation in policy making. Several pitfalls beset such an approach. Scientists do not solely value "truth." Indeed, many value political power and influence over policy formation. Or a scientist may use or exchange values of skill or enlightenment for values of power and wealth. The transfer of knowledge into power occurs as normally and easily as a Harriman, Rockefeller, or Kennedy parleying wealth into political power. Science, truth, and knowledge are often tools for the pursuit of political power, just like opinion polls, political debts, and promises of support. But even if scientists' "truth" is expected to check and balance politicians' "power," we have no guarantee or evidence that truth will not overpower power, or power overpower truth, in the long run.

Elites do not always check and balance each other. They may be as mutually interdependent as they are mutually distrustful. One elite may fulfill functions like goal attainment and adaptation for society. Another may promote pattern maintenance and integration. But both are required to make the society function and survive. Thus both elites have an interest in each others' activities. Functions are more likely to coexist with than counterbalance other functions. Therefore, scientists may have an interest in the system's survival, may gain from its growth, and may benefit from a strong state that will provide stability and insulate scientists from the excesses of the common man. Or scientists may voluntarily join or be absorbed within innumerable *complexes* or *amalgams,* such as the *security state, warfare state, welfare state, contract state, educational state,* and *industrial state.* The various groups in a complex achieve their own particular objectives and the complex's objectives through cooperation and symbiosis. They slap each others' backs and roll each others' logs.

Perhaps no elite can check another elite. Perhaps a scientific elite cannot be expected to check political and economic elites. Kenneth Boulding, in a book review criticizing arguments relying on science to check these political and economic elites, argues that:

> Goliaths will always hang together; it is only David with a sling-shot who can bring them down. Both Galbraith *[The New Industrial State]* and Nieburg *[In the Name of Science]* are relying on Goliaths to tame each other and this may be unrealistic.[15]

For Boulding, only "outsiders" such as religious elites or the church can withhold legitimacy from the state or challenge the system.[16] Insiders, like scientists and the universities, cannot serve as countervailing powers.

Though there are serious difficulties associated with relying on scientists and leaders balancing one another's interests and objectives, no realistic alternative is available. Scientists now are variously a powerful elite among elites, an interest group, a skilled apolitical elite, or a politically fragile and fragmented group.[17] Their status varies with the issue and policy. Political figures "use" scientists for political ends.[18] And scientists "use" political figures for scientists' ends. They may not always check or distrust one another, but the possibility is maximized the more they are in confrontation.

Therefore, there may be wisdom in placing scientists alongside political leaders in positions of responsibility.[19] Leaders would develop more sophistication in understanding science and scientists and would learn how to seize the initiative from scientists. Scientists would develop an appreciation for the demands of politics. Scientists' conflicts of interest might be eliminated or reduced. Better coordination and planning for science and the use of science might emerge. Scientists themselves would have operational and some political responsibilities. And the political climate would set its usual limits of consensus built around the attitudes of politically significant groups.

SCIENTISTS AND SCIENCE:
A NATION'S TEMPTATION

> The heaviest responsibility of the scientist to society may be to refuse to make himself useful.
> Scott Buchanan in *Science, Scientists and Politics* (Center for the Study of Democratic Institutions)[20]

Man's greatest difficulty lies not in his control over the scientist's behavior but in his lack of control over himself. He only very erratically controls his own destiny. Democratic majorities are often unable to check and restrain themselves. But both man and majorities have been relatively successful in shaping and controlling scientists' participation in political affairs. Society rather than science has proved the more powerful partner through the 1960s.

But the scientist needs man—as patron, customer, and subject matter. The interdependence of scientist and citizen is an established fact. Still, not everything man "needs" from scientists will be automatically "good" for him. Therefore, as Scott Buchanan dramtically contends, the scientist may bear a special burden. And yet, historically, though scientists might often have withheld the Brechtian "corn" of truth or concealed knowledge, more often than not their fruits have shaped the lives of men and policies of governments. Still, the moral question for every scientist must remain. Should he deliberately not develop new knowledge? Should he conceal his findings? Should he serve societies' needs and desires?

The analysis thus far of scientists' influence in policy making provides one major insight. Their influence and participation is shaped significantly by political forces and interests. Legislators, special interests, political executives, and the public all set forth the objectives and courses of public policies. Scientists may tempt these people with possibilities and knowledge, but the decision to forge and pursue a policy remains with these people alone. Scientists have rarely foisted anything on society that its policy-makers, citizens, or politically significant groups did not feel they wanted or did not encourage.

I would not, however, deny that scientists often lead society into new visions and desires. Nor would I deny that scientists may argue for their own political preferences, create demands for their products and services, and attempt to influence public policies. The space program and weapons development are clear examples. So is the whole field of science policy making. Scientists do pursue power and influence over policy within the inner councils of policy making and through the public forum.

Likewise, it is clear that these men of truth and knowledge are often allied with men of power, each pursuing their respective ends of truth and power. Scientists who pursue a pragmatic type of "truth" influence weapons policy making. Weapons that work are "true" and scientifically correct, but they concurrently serve the interests of political leaders who want power. Truth will not always check power, whether it comes from the scientist, philosopher, or theologian. But I also argue that scientists' skills and views of the world often coincide with the needs and views of some policy-makers, special interests, or publics.

So the difficulties raised are part and parcel of a larger dilemma. Can a democratic majority, special interest, or political leadership police itself? Who checks the majority?

Policy-makers, special interests, and citizens can only control science and their use of scientists by controlling themselves. They must learn to assess and anticipate the effects of new technologies. They must learn when to use and when not to use the skills and fruits of scientists' labor. They must exercise control over their own needs, using science for some needs like health and perhaps less uncritically using science for other needs like weapons for defense.

Men need weapons scientists and "strategy scientists" because they cannot seem to control international affairs in other ways. They need weapons, economists, and social scientists because they cannot seem to control their domestic affairs in other ways. Too often we manifest an uncritical craving and deference to scientific and technological solutions to human problems. Military means are not always the only solution to international conflicts or domestic rioting. Tranquilizers are not the only avenue to mental health. Some needs are often unreal or imagined, but the demands they produce for science are nevertheless powerful and effective. And some needs, including demands for weapons and sociological diagnoses, are necessary. But in many cases, other solutions are available if only men saw their problems in different terms.

Victor Ferkiss, in a recent and stimulating book, *Technological Man: The Myth and Reality*, argues that man must become "technological man." If he does not, Ferkiss suggests, man will continue to be *controlled by* technology rather than *controlling* technology.[21] Survival in a technological age demands that we transcend our lingering status as *industrial men* and *liberal democratic men* and develop a capacity to cope with and use technology in a critical fashion.

Men need science, and scientists need men. So much of scientists' influence stems from (1) societies' need for scientists' skills and knowledge; (2) the collective fears, dissatisfactions, and insecurity that render and lead peoples to defer unnecessarily to scientists' judgment and policy positions; and (3) men's inability to resist scientific and technological feasibilities.

Science enables government to distribute desired values and fulfill politically powerful demands. Science provides a means to military superiority and communal security in the fields of health, weapons, defense, and weather policy. We have forged an alliance of mutual benefit between scientist and society. And society is the more influential and powerful partner determining when it will permit science and scientists to shape the affairs and lives of men.

Science and scientists have emerged from the traditions of rationalism and positivism. They provide *technical rationality* for a society, giving it the ability to predict and control its future on the basis of empirical knowledge.[22] Through experiment, logic, and reason, man can describe and explain his environment and self with some measure of certainty.

Auguste Comte, as a positivist, argued that his "social physics" would enable man to discover laws governing his own behavior, which he could then manipulate to establish an ordered, tranquil social system.[23] Comte's utopian technocracy in his *Système de Politique Positive* would supplant traditional religion with the guidance of captains of industry, inventors, and scientists. Man would allow science and technology to shape his life. Neither Catholicism nor democratic majorities nor publics would determine society's course. Man's problems would be solved rationally by experts. Scientists would provide man with "self-consciousness" and direction for policies.

Scientists, methodologically averse to the bargaining, incrementalism, majorities, and compromise of political life, would transcend that form of policy making. Social problems would be solved synoptically and prior to their development. Reaction and incrementalism would become social anachronisms.

The tension remains. Should we make policy this way? Where should a society obtain its "self-consciousness"? Philosophers and social thinkers might provide "self-consciousness," [24] and so might social and physical scientists. But so might nonscientists and nonphilosophers comprising vast publics and democratic majorities. So might, for that matter, any individual serve as the source of his own awareness of self and society.

Whom shall an individual trust to develop his consciousness of self and forge the course of his society? We would agree that most societies have and might want to increase their use of what Amitai Etzioni calls "knowledge units." [25] But to what extent should societies and majorities defer to the diagnoses, "truths," and policies put forth by an expert few?

Yehezkel Dror feels that:

> Conscious calculation of social direction must therefore partly replace the automatic and semispontaneous adjustment of society to new knowledge that generally sufficed in the past.[26]

But how will society use the scientists and men of reason who might "consciously calculate" its direction? Will we use them merely to feed our appetites and unchecked desires? Or will we put them to work developing new ways of making policy and social decisions?

Until now, we have used scientists more for their "policy issue knowledge" than for their "policy making knowledge." This is perhaps because the potential benefits were more readily apparent.[27] But "policy making knowledge" or "metapolicy" such as specific deference to operations research, systems analysis, policy sciences, organizational analysis, and game theory can tell us much more and lend a higher level of consciousness. Men are reluctant to change what they do but even more reluctant to change their ways of changing. Thus what Dror calls *policy issue knowledge* (medical sciences on health policy, social sciences on segregation, etc.) has been used long before we consciously adopted *policy-making knowledge* to further increase our society's "self-consciousness."

Perhaps society has decided to derive its "self-consciousness" from scientists. Perhaps philosophers and social thinkers, intuition, and the unconscious will no longer be such influential sources of understanding of ourselves as they once were. And we will turn to the modern heirs of rationalism and positivism.

But in turning to scientists, we run the danger of succumbing to a modern day Scylla and Charybdis. Society could easily bind scientists so tightly to its demands that they could not flourish. Or society could gluttonously grab at scientists' products as if each were a matter of survival.

Society must be careful in using scientists for its ends. Too much control and constraint can blunt scientists' creative spirit and malign science's internal development. We might even destroy the possibility that scientists could provide us with diagnosis and analysis for debates and technologies for material progress. Fewer "right" answers might emerge from science. We might reduce our supply of new technological, scientific, and policy-making options.

Scientists have been a "liberal" force in society. We might destroy that contribution. They have expanded policy-makers' awareness of situations and the environment. They have exposed the illnesses and psychological deformities of the "disadvantaged" minority. They have helped government meet the material needs of minorities and the common man. Bureaucracies, stagnated by age, have been given some measure of flexibility to adjust and respond to a changing society. Americans' attention has been shifted from the *property rights* of polluters to the *environmental rights* of consumers of polluted air, water, and foods. We cannot afford to diminish this great contribution of science in enabling the spread of material and human welfare to all mankind. For in this role scientists have been a force for "democratic" and "liberal" change.

Scientists' contribution of reason and self-consciousness to a society enables that society to understand and guide itself as an autonomous unit. Such a capacity is probably essential to a social system's survival. But if man cannot temper his seemingly insatiable needs and use scientists' knowledge with critical judgment, then perhaps "the heaviest responsibility of the scientist to society may be to refuse to make himself useful." [28]

Thus Bertolt Brecht's Cardinal Barberini may have had a wise admonition in that "a prudent man concealeth knowledge." Galileo might well have withheld the "corn"—his knowledge of the nature of the universe.[29] But then the man of reason, rationalism, and positivism would have made a decision perhaps not properly his. Or suppose he did "tell it like it is." Either way, as blind men or wise men, societies and democratic majorities would be left alone to sow the seeds and reap the fruits of their own destruction.

Notes

1. Frederic N. Cleaveland, *Science and State Government* (Chapel Hill: University of North Carolina Press, 1959).

2. *Ibid.*, p. 153.

3. Harvey M. Sapolsky, "Science Advice for State and Local Governments," *Science*, 160 (April 19, 1968), pp. 280–284. Sapolsky's article and discussion of factors influencing states' and cities' use of science and support for science is a pioneering effort, standing alongside Cleaveland's book in this respect.

4. Richard C. Cortner, "Strategies and Tactics of Litigants in Constitutional Cases," *Journal of Public Law*, 17 : 2 (1968), pp. 287–307. See particularly pp. 305–306.

5. Useful analogies may be drawn from studies on the role of scientists and science in developing *countries*. Such nations need basic research, which cannot produce quick returns and which cannot really be always imported or borrowed. Science is alien to their traditional, bureaucratic, or ideology dominated cultures. No indigenous demands exist to support research, even in agriculture. There are no native scientists or facilities for training scientists. National planners have no use for scientific contributions to their efforts. There is little scientific exchange or communication with other countries. See Stevan Dedijer, "Underdeveloped Science in Underdeveloped Countries," *Minerva*, II : 1 (Autumn, 1963), pp. 61–81, and Michael J. Moravcsik, "Some Practical Suggestions for the Improvement of Science in Developing Countries," *Minerva*, IV : 3 (Spring, 1966), pp. 381–390.

Many problems and opportunities beset such developing lands. They include population control, resource development, fuels, energy sources, dietary needs, water sources, new food sources, construction of shelter and manufacturing facilities, and urbanization. See Richard L. Meier, *Science and Economic Development* (Cambridge: M.I.T. Press, 1966).

David Apter, however, suggests that science should dovetail quite well with the planning "ideal." Indeed, science provides a new basis of legitimacy in modernization—the rationality of social engineering. Scientists, though they may join with other elites, will nevertheless find themselves at odds with the more traditional, propertied elites. And science may harm the position of established ideologies that bind and solidify a nation, since science's authority inheres in knowledge and loyalty goes first to a profession. Science itself becomes an ideology (rationalism, positivism, reason, rationality, empiricism) which may variously complement and conflict with nationalism or socialism as ideologies. Apter, *The Politics of Modernization* (Chicago: University of Chicago Press, 1965), pp. 326–327, 343–356, 434–450.

6. This discussion of *behavioral and physical technologies* is drawn, some of it *verbatim*, from my paper, "Political Arenas and Contributions of Physical and Behavioral Sciences and Technologies to Policymaking" (Paper Delivered at the Sixty-Fifth Annual Meeting of the American Political Science Association, Commodore Hotel, New York, September 2–6, 1969), revised as "Political Arenas, Life Styles, and the Contribution of Technologies to Policymaking," *Policy Sciences*, I : 3 (tentative, Fall 1970).

7. Some basic sources in the sociology of science are Bernard Barber, *Science and the Social Order* (London: George Allen and Unwin, Ltd., 1953); Bernard Barber and Walter Hirsch (eds.), *The Sociology of Science* (New York: Free Press, 1962); John D. Bernal, *The Social Function of Science* (New York: Macmillan, 1939); Lewis A. Coser, *Men of Ideas* (New York: Free Press, 1965); Gerard DeGré *Science as a Social Institution* (New York: Random House, 1955); Lewis S. Feuer, *The Scientific Intellectual* (New York: Basic Books, 1963); Norman Kaplan (ed.), *Science and Society* (Chicago: Rand McNally, 1965); Thomas S. Kuhn, *The Structure of Scientific Revolutions* (Chicago: University of Chicago Press, 1962); Norman W. Storer, *The Social System of Science* (New York: Holt, Rinehart and Winston, 1966); Derek J. de Solla Price, *Science Since Babylon* (New Haven: Yale University Press, 1961); Robert K. Merton, "Science, Technology and Society in Seventeenth Century England," *Osiris*, IV (1938), pp. 360–362; and Joseph Ben-David, "The Scientific Role: The Conditions of its Establishment in Europe," *Minerva*, IV : 1 (Autumn, 1965), pp. 15–54.

8. See Albert Parry, *The New Class Divided: Russian Science and Technology versus Communism* (New York: Macmillan, 1966); David Joravsky, *Soviet Marxism and Natural Science, 1917–1932* (London: Routledge, Kegan and Paul, 1961); and George Fisher (ed.), *Science and Ideology in Soviet Society* (New York: Atherton Press, 1967).

9. Joseph W. Eaton, "Symbolic and Substantive Evaluative Research," *Administrative Science Quarterly*, VI (March, 1962), pp. 421–442.

10. Ashley L. Schiff, *Fire and Water: Scientific Heresy in the Forest Service* (Cambridge: Harvard University Press, 1962).

11. Walter Goldstein, "The Science Establishment and Its Political Control," *Virginia Quarterly Review*, 43 : 3 (Summer, 1967), pp. 343-371.

12. Avery Leiserson, "Scientists and the Policy Process," *American Political Science Review*, LIX : 2 (June, 1965), p. 416.

13. James Ward Smith, *A Theme for Reason* (Princeton: Princeton University Press, 1957), pp. 110-112.

14. Alfred de Grazia, "The Politics of Science and Dr. Velikovsky," *American Behavioral Scientist*, VII : 1 (September, 1963), pp. 3-68. Professor de Grazia both edits and contributes to this special supplement. Alfred de Grazia, "A Concept of Scientists and their Organization," *American Behavioral Scientist*, VI : 4 (December, 1962), pp. 30-34, de Grazia argues that "power" and "dogmatic" models of scientists' internal group relationships explain much of their behavior. Leaders of science, concerned with the prestige, salaries, and positions they have attained, use power to retain those benefits. For many, their prestige and salaries rest on "truths" they discovered and represent. So challenges to those "truths" from younger scientists or a Dr. Velikovsky may be met with power and assertions of the old dogma or "truths." This picture differs greatly from the "rational" model of science that most of us hold as a stereotype. But it does indicate that scientists are men of power as well as men of truth. And any efforts at integrating scientists into policy making must cope with this desire of scientists.

Renato Taguiri provides some additional evidence of scientists' value for power. Polling religious figures, research managers, executives, and scientists on their preferred values, Taguiri found that scientists did indeed rank theoretical values first (truth; rational, critical, and empirical activities). However, they ranked political values (power) second, *higher than executives*, research managers, and religious figures ranked political values. Taguiri, "Value Orientations and the Relationship of Managers and Scientists," *Administrative Science Quarterly*, X : 1 (June, 1965), pp. 39-51.

15. Kenneth Boulding, "The Scientific-Military-Industrial Complex," *Virginia Quarterly Review*, 43 : 4 (Autumn, 1967), p. 679. I have placed the books' titles in brackets since the quotation comes from a book review.

16. I doubt that the church as another institution can meet the need, for often it too seems to be an "insider."

17. Authors with differing viewpoints include Ralph E. Lapp, *The New Priesthood* (New York: Harper and Row, 1965); Jacques Ellul, *The Technological Society* (New York: Alfred A. Knopf, 1964); Don

K. Price, *The Scientific Estate* (Cambridge: Harvard University Press, 1965); Robert C. Wood, "Scientists and Politics: The Rise of an Apolitical Elite," *Scientists and National Policy-Making*, ed. by Robert Gilpin and Christopher Wright (New York: Columbia University Press, 1964); and Daniel S. Greenberg, "The Myth of the Scientific Elite," *Public Interest*, I : 1 (Fall, 1965), pp. 51–62.

18. Harvey Wheeler's account of a consultantship on state governmental reorganization illustrates how politicians can *use* scientists. Wheeler, "The Short and Happy Life of a Research Consultantship," *Western Political Quarterly*, 13 (September, 1960), pp. 852-857. Anthony Downs' article on the uses of economic advice is also indicative. Downs, "Some Thoughts on Giving People Economic Advice," *American Behavioral Scientist*, IX : 1 (September, 1965), pp. 30–32.

19. Robert T. Golembiewski, *Organizing Men and Power: Patterns of Behavior and Line-Staff Models* (Chicago: Rand McNally, Inc., 1967). Golembiewski suggests placing experts within the chain of command, not outside as older line-staff, alter ego, and "neutral and inferior instrument" theories would have done. Scientists would be integrated into bargaining, policy responsibility, and operational activities. They would compete on a par with lawyers, bureaucrats, and political executives. Tension and stress would be reduced between expert and generalist. Cooperation would be essential, and some balance of technical and political criteria might be achieved in policies. Scientists would have to confront and adapt to budgetary restraints, public inertia, and the problems of dealing with legislatures. But perhaps more importantly, political leaders would also be integrated into agencies and bureaus long dominated by scientists.

20. Center for the Study of Democratic Institutions, *Science, Scientists and Politics* (New York: Fund for the Republic, 1963).

21. Victor Ferkiss, *Technological Man: The Myth and Reality* (New York: George Braziller, 1969).

22. Paul Diesing has distinguished *technical rationality* from legal, social, political, and economic rationality. As sciences that diagnose and predict events, even the behavioral sciences involve technical (means-ends) rationality. But their subject matter lies in the realm of social (socialization, adjustment), political (compromise), and economic (market bargaining) rationalities. Thus social scientists would be equipped to understand the role of these forms of rationality in societies. Diesing, *Reason in Society* (Urbana: University of Illinois Press, 1962).

23. Auguste Comte, *A General View of Positivism*, trans. by J. H. Bridges (Stanford, California: Academic Reprints, n.d.). See Comte, *Cours de Philosophie Positive* in six volumes from 1830–1842 and *Système de Politique Positive* in four volumes from 1851–1854.

24. H. Stuart Hughes, *Consciousness and Society* (New York:

Vintage Random, 1958), p. 3. The term "self-consciousness" is Hughes' term.

25. Amitai Etzioni, *The Active Society* (New York: Free Press, 1968).

26. Yehezkel Dror, *Public Policymaking Reexamined* (San Francisco: Chandler Publishing Co., 1968), p. 5.

27. Consult Dror, *op. cit.*, pp. 1-10, 220-245, 267-278, and Appendix C.

28. Scott Buchanan, *Science, Scientists and Politics* (New York: Fund for the Republic, Center for the Study of Democratic Institutions, 1963). Buchanan's contribution to the pamphlet is brief but incisive.

29. Bertolt Brecht, *Galileo* (New York: Grove Press, 1966), p. 77.

Bibliography

This bibliography concentrates on the political aspects of scientists' relationship to public policy. Since many comprehensive bibliographies already permeate the field of science and public policy, this listing provides only a selected bibliography for readers with limited and immediate interests. However, though these existing bibliographies differ from this list in perspective and comprehensiveness, they are invaluable to any extensive inquiry.

BIBLIOGRAPHIES

Caldwell, Lynton K. (ed.). *Science, Technology, and Public Policy: A Selected and Annotated Bibliography*. Prepared for the National Science Foundation by the Program in Public Policy for Science and Technology, Department of Government, Indiana University, Bloomington. Volume I: Books, Monographs, Government Documents, and Journal Issues. Volume II: Journal Articles and Addenda to Volume I. An *excellent* bibliography.

Battelle Memorial Institute (Columbus Laboratories). *Science Policy Bulletin*. Bimonthly since October, 1967, and mailed free upon request. Current and annotated.

Universal Reference System. *Public Policy and the Management of Science*. Volume IX in the Codex Series in Political Science. Princeton: Princeton Research Publishing Company, 1969. Consists of annotated initial Codex, annual cumulative additions, and current supplements, and available in libraries or by subscription.

Accessions Lists, Information Center, Harvard University Program on Technology and Society, Cambridge, Massachusetts. Published periodically in Xerox form. Annotated and indexed.

Sapolsky, Harvey, and Daniel Rich. *Science and Public Policy: An Introductory Bibliography.* Cambridge: Department of Political Science, M.I.T., 1967. Limited but basic listing.

Program in Science and Public Policy, Department of Political Science, Purdue University, Lafayette, Indiana. Unannotated, brief but diverse list (1967).

PUBLIC DOCUMENTS

U.S. House of Representatives, Research and Technical Programs Subcommittee, Committee on Government Operations. *Federal Research and Development Programs: The Decisionmaking Process.* Staff Report. 89th Cong., 2nd Sess., January 7–11, 1966.

U.S. House of Representatives, Research and Technical Programs Subcommittee, Committee on Government Operations. *The Use of Social Research in Federal Domestic Programs.* Volumes I–IV, 90th Cong., 1st Sess., April, 1967.

U.S. Senate, Subcommittee of National Policy Machinery, Committee on Government Operations. *Organizing for National Security: Science, Technology, and the Policy Process.* 86th Cong., 2nd Sess., April 25–27, 1960.

BOOKS

Art, Robert J. *The TFX Decision: McNamara and the Military.* Boston: Little, Brown and Company, 1968.

Armine, Michael. *The Great Decision: The Secret History of the Atomic Bomb.* New York: G. P. Putnam's Sons, 1959.

Armytage, W. H. G. *The Rise of the Technocrats: A Social History.* London: Routledge and Kegan Paul, 1965.

Bacon, Francis. *New Atlantis.* Contained in *Ideal Commonwealths.* Edited by Henry Morley. New York: E. P. Dutton and Co., 1885.

Bailey, Stephen K. *Congress Makes a Law: The Story Behind the Employment Act of 1946.* New York: Columbia University Press, 1950.

Barber, Bernard. *Science and the Social Order.* London: George Allen and Unwin, Ltd., 1953.

Barber, Bernard, and Hirsch, Walter (eds.). *The Sociology of Science.* New York: Free Press, 1962.

Barber, Richard J. *The Politics of Research.* Washington: Public Affairs Press, 1966.

Baritz, Loren. *The Servants of Power: A History of the Use of Social Science in American Industry.* Middletown, Connecticut: Wesleyan University Press, 1960.

Batchelder, Robert C. *The Irreversible Decision, 1939-1950.* New York: Macmillan Company, 1961.

Bauer, Raymond A. (ed.) *Social Indicators.* Cambridge: M.I.T. Press, 1966.

Bauer, Raymond, Pool, Ithiel de Sola, and Dexter, Lewis A. *American Business and Public Policy.* New York: Atherton Press, 1964.

Bauer, Raymond and Gergen, Kenneth (eds.). *The Study of Policy Formation.* New York: Free Press, 1968.

Benoit, Emile and Boulding, Kenneth E. (eds.). *Disarmament and the Economy.* New York: Harper and Row, 1963.

Bernal, John Desmond. *The Social Function of Science.* New York: Macmillan Company, 1939.

Bottomore, T. B. *Elites and Society.* New York: Basic Books, 1964.

Bottomore, T. B. and Rubel, Maximilien (eds.). *Karl Marx: Selected Writings in Sociology and Social Philosophy.* New York: McGraw-Hill, 1964.

Boulding, Kenneth E. *The Impact of the Social Sciences*. New Brunswick, New Jersey: Rutgers University Press, 1966.

Braybrooke, David and Lindblom, Charles E. *A Strategy of Decision: Policy Evaluation as a Social Process*. New York: Free Press, 1963.

Brecht, Bertolt. *Galileo*. Edited by Eric Bentley, New York: Grove Press, 1966.

Brodie, Bernard. *The American Scientific Strategists*. Santa Monica, California: The Rand Corporation, 1964.

Burke, John G. (ed.). *The New Technology and Human Values*. Belmont, California: Wadsworth Publishing Co., 1966.

Burnham, James. *The Managerial Revolution*. New York: John Day Co., 1941.

Campanella, Thomas. *The City of the Sun*. Contained in *Ideal Commonwealths*. Edited by Henry Morley. New York: E. P. Dutton and Co., 1885.

Center for the Study of Democratic Institutions. *Science, Scientists, and Politics*. New York: Fund for the Republic, 1963.

Clark, John J. *The New Economics of National Defense*. New York: Random House, 1966.

Cleaveland, Frederic N. *Science and State Government*. Chapel Hill: University of North Carolina Press, 1959.

Commager, Henry Steele (ed.). *Lester Ward and the Welfare State*. Indianapolis: Bobbs-Merrill Company, Inc., 1967.

Cook, Fred J. *The Warfare State*. New York: Macmillan Company, 1962.

Coser, Lewis A. *Men of Ideas: A Sociologist's View*. New York: Free Press, 1965.

Cox, Donald W. *America's New Policy Makers: The Scientists' Rise to Power*. Philadelphia: Chilton Co., 1964.

Danielson, Michael. *Federal-Metropolitan Politics and the Commuter Crisis*. New York: Columbia University Press, 1965.

Davies, J. Clarence. *The Politics of Pollution.* New York: Pegasus, 1970.

DeGre, Gerard. *Science as a Social Institution.* New York: Random House, 1955.

Dewey, John. *The Public and Its Problems.* New York: Henry Holt and Company, 1927.

Downs, Anthony. *Inside Bureaucracy.* Boston: Little, Brown and Company, 1967.

Dror, Yehezkel. *Public Policymaking Reexamined.* San Francisco: Chandler Publishing Co., 1968.

Dupre, Stefan J. and Lakoff, Sanford A. *Science and the Nation: Policy and Politics.* Englewood Cliffs: Prentice Hall, 1962.

Dupree, A. Hunter (ed.). *Science and the Emergence of Modern America.* Chicago: Rand McNally and Company, 1963.

Dupree, A. Hunter. *Science in the Federal Government: A History of Policies and Activities to 1940.* Cambridge: Belknap Press of the Harvard University Press, 1957.

Duscha, Julius. *Arms, Money and Politics.* New York: Ives Washburn, Inc., 1965.

Easton, David. *A Systems Analysis of Political Life.* New York: John Wiley and Sons, Inc., 1965.

Edelman, Murray. *The Symbolic Uses of Politics.* Urbana: University of Illinois Press, 1967.

Eiduson, Bernice T. *Scientists: Their Psychological World.* New York: Basic Books, 1962.

Elder, Robert T. *The Policy Machine: The Department of State and American Foreign Policy.* Syracuse: Syracuse University Press, 1960.

Ellul, Jacques. *The Technological Society.* New York: Alfred A. Knopf, 1964.

Engler, Robert. *The Politics of Oil.* Chicago: University of Chicago Press, 1961.

Etzioni, Amitai. *The Active Society*. New York: Free Press, 1968.

Etzioni, Amitai. *The Moon-Doggle: Domestic and International Implications of the Space Race*. New York: Doubleday and Co., Inc., 1964.

Fainsod, Merle, Gordon, Lincoln, and Palamountain, Joseph C., Jr. *Government and the American Economy*. New York: W. W. Norton and Company, Inc., 1959.

Feis, Herbert. *Japan Subdued: The Atomic Bomb and the End of the War in the Pacific*. Princeton: Princeton University Press, 1961.

Ferkiss, Victor. *Technological Man*. New York: George Braziller, 1969.

Feuer, Lewis S. *The Scientific Intellectual: The Psychological and Sociological Origins of Modern Science*. New York: Basic Books, 1963.

Fisher, George. *Science and Politics: The New Sociology in the Soviet Union*. Ithaca: Cornell University Center for International Studies, 1964.

Fisher, George (ed.). *Science and Ideology in the Soviet Union*. New York: Atherton Press, 1967.

Flash, Edward S., Jr. *Economic Advice and Presidential Leadership: The Council of Economic Advisers*. New York: Columbia University Press, 1965.

Fritschler, A. Lee. *Smoking and Politics*. New York: Appleton-Century-Crofts, 1969.

Galbraith, John Kenneth. *The New Industrial State*. Boston: Houghton Mifflin Company, 1967.

Gilpin, Robert. *American Scientists and Nuclear Weapons Policy*. Princeton: Princeton University Press, 1962.

Glock, Charles Y., *et al*. *Case Studies in Bringing Behavioral Science into Use*. Volume I: *Studies in the Utilization of Behavioral Science*. Stanford: Institute for Communication Research, Stanford University, 1961.

Goldman, Marshall I. *Controlling Pollution: The Economics of a Cleaner America*. Englewood Cliffs: Prentice Hall, 1967.

Golembiewski, Robert T. *Organizing Men and Power: Patterns of Behavior and Line-Staff Models*. Chicago: Rand McNally, Inc., 1967.

Green, Harold P. and Rosenthal, Alan. *Government and the Atom*. New York: Atherton Press, 1963.

Greenberg, Daniel S. *The Politics of Pure Science*. New York: The New American Library, 1967.

Griffith, Alison. *The National Aeronautics and Space Act: A Study of the Development of Public Policy*. Washington: Public Affairs Press, 1962.

Groueff, Stephane. *Manhattan Project*. Boston: Little, Brown and Company, 1967.

Hagstrom, Warren O. *The Scientific Community*. New York: Basic Books, 1965.

Hammond, Paul Y. *Organizing for Defense*. Princeton: Princeton University Press, 1961.

Hardin, Charles M. *The Politics of Agriculture*. Glencoe, Illinois: Free Press, 1952.

Harrington, Michael. *The Accidental Century*. Baltimore: Penguin Books, 1965.

Helmer, Olaf. *Social Technology*. New York: Basic Books, 1966.

Herzog, Arthur. *The War-Peace Establishment*. New York: Harper and Row, 1965.

Hewlett, Richard G. and Anderson, Oscar E., Jr. *The New World, 1939–1946*. Volume I. University Park: Pennsylvania State University Press, 1962.

Hilsman, Roger. *Strategic Intelligence and National Decisions*. Glencoe, Illinois: Free Press, 1956.

Hitch, Charles J. and McKean, Roland N. *The Economics of Defense in the Nuclear Age*. Cambridge: Harvard University Press, 1960.

Horowitz, Irving Louis (ed.). *The Rise and Fall of Project Camelot: Studies in the Relationship Between Social Science and Practical Politics*. Cambridge: M.I.T. Press, 1967.

Hughes, H. Stuart. *Consciousness and Society*. New York: Random House, 1958.

Huntington, Samuel P. *The Common Defense*. New York: Columbia University Press, 1961.

Jacobson, Harold K. and Stein, Eric. *Diplomats, Scientists, and Politicians: The United States and the Nuclear Test Ban Negotiations*. Ann Harbor: University of Michigan Press, 1966.

Janowitz, Morris. *Sociology and the Military Establishment*. New York: Russell Sage Foundation, 1959.

Joravsky, David. *Soviet Marxism and Natural Science, 1917–1933*. New York: Columbia University Press, 1961.

Jungk, Robert. *Brighter Than a Thousand Suns: The Moral and Political History of the Atomic Scientists*. London: Victor Gollancz, Ltd., 1958.

Kaplan, Norman (ed.). *Science and Society*. Chicago: Rand McNally, 1965.

Kauffman, W. W. *The McNamara Strategy*. New York: Harper and Row, 1964.

Keller, Suzanne. *Beyond the Ruling Class: Strategic Elites in Modern Society*. New York: Random House, 1963.

Kirkendall, Richard S. *Social Scientists and Farm Politics in the Age of Roosevelt*. Columbia: University of Missouri Press, 1966.

Kuhn, Thomas S. *The Structure of Scientific Revolutions*. Chicago: University of Chicago Press, 1962.

Lakoff, Sanford A. (ed.). *Knowledge and Power: Essays on Science and Government*. New York: Free Press, 1966.

Lapp, Ralph E. *The New Priesthood*. New York: Harper and Row, 1965.

Lapp, Ralph E. *The Weapons Culture*. W. W. Norton and Co., 1968.

Lasswell, Harold D. *The Analysis of Political Behavior*. New York: Oxford University Press, 1947.

Lasswell, Harold D. *The Decision Process: Seven Categories of Functional Analysis*. College Park: College of Business and Public Administration, Bureau of Governmental Research, University of Maryland, 1956.

Lazarsfeld, Paul F., Sewall, William H., and Wilensky, Harold L. (eds.). *The Uses of Sociology*. New York: Basic Books, 1967.

Levine, Robert A. *The Arms Debate*. Cambridge: Harvard University Press, 1963.

Lindblom, Charles E. *The Intelligence of Democracy: Decision Making through Mutual Adjustment*. New York: Free Press, 1965.

Lovejoy, Wallace F. and Homan, Paul T. *Economic Aspects of Oil Conservation Regulation*. Baltimore: Johns Hopkins Press, 1967.

Manuel, Frank E. *The New World of Henri Saint-Simon*. Cambridge: Harvard University Press, 1956.

McCamy, James L. *Science and Public Administration*. Birmingham: University of Alabama Press, 1960.

Merton, Robert K. *Social Theory and Social Structure*. Glencoe, Illinois: Free Press, 1949.

Millis, Walter, Mansfield, Harvey C., and Stein, Harold. *Arms and the State*. New York: Twentieth Century Fund, 1958.

Mills, C. Wright. *The Power Elite*. New York: Oxford University Press, 1959.

Newman, James R. and Miller, Byron S. *The Control of Atomic Energy: A Study of Its Social, Economic and Political Implications*. New York: McGraw-Hill Book Company, Inc., 1948.

Nieburg, Harold L. *In the Name of Science*. Chicago: Quadrangle Books, 1966.

Bibliography

Nieburg, Harold L. *Nuclear Secrecy and Foreign Policy.* Washington: Public Affairs Press, 1964.

Nomad, Max. *Aspects of Revolt.* New York: Bookman Associates, 1959.

O'Leary, Michael Kent. *The Politics of American Foreign Aid.* New York: Atherton Press, 1967.

Organization for Economic Cooperation and Development. *Science and the Policies of Governments.* Paris: OECD, 1963.

Parry, Albert. *The New Class Divided.* New York: Macmillan Company, 1966.

Penick, James L., Jr. (ed.). *The Politics of American Science: 1939 to the Present.* Chicago: Rand McNally, Inc., 1965.

Plato, *The Republic and Other Works.* Translated by B. Jowett. Garden City, New York: Doubleday and Company, Inc., 1960.

Polenberg, Richard. *Reorganizing Roosevelt's Government: The Controversy over Executive Reorganization, 1936–1939.* Cambridge: Harvard University Press, 1966.

Price, Don K. *Government and Science.* New York: New York University Press, 1954.

Price, Don K. *The Scientific Estate.* Cambridge: Harvard University Press, 1965.

Rainwater, Lee and Yancey, William L. (eds.). *The Moynihan Report and the Politics of Controversy.* Cambridge: M.I.T. Press, 1967.

Ranney, Austin (ed.). *Political Science and Public Policy.* Chicago: Markham Publishing Co., 1968.

Roe, Anne. *The Making of a Scientist.* New York: Dodd, Mead and Company, 1953.

Schelling, Thomas C. *The Strategy of Conflict.* Cambridge: Harvard University Press, 1963.

Schelling, Thomas C. and Halperin, Morton H. *Strategy and Arms Control.* Twentieth Century Fund, 1961.

Schiff, Ashley L. *Fire and Water: Scientific Heresy in the Forest Service.* Cambridge: Harvard University Press, 1962.

Schilling, Warner R., Hammond, Paul Y., and Snyder, Glenn H. *Strategy, Politics, and Defense Budgets.* New York: Columbia University Press, 1962.

Science Policy Research Division, Library of Congress. *Technical Information for Congress.* (91st Cong., 1st Sess., Report for the Subcommittee on Science, Research, and Development, House of Representatives) Washington: Government Printing Office, 1969.

Silverman, Corinne. *The President's Economic Advisers.* University: University of Alabama Press, 1959.

Skolnikoff, Eugene B. *Science, Technology, and American Foreign Policy.* Cambridge: M.I.T. Press, 1967.

Smerk, George M. *Urban Transportation: The Federal Role.* Bloomington: Indiana University Press, 1965.

Smith, Alice Kimball. *A Peril and a Hope: The Scientists' Movement in America, 1945–1947.* Chicago: University of Chicago Press, 1965.

Smith, Bruce L. R. *The RAND Corporation.* Cambridge: Harvard University Press, 1966.

Smith, Frank E. *The Politics of Conservation.* New York: Pantheon House, 1966.

Snow, C. P. *Science and Government.* New York: Mentor, New American Library, 1962.

Spanier, John W. and Nogee, Joseph L. *The Politics of Disarmament.* New York: Frederick A. Praeger, 1962.

Stein, Herbert. *The Fiscal Revolution in America.* Chicago: University of Chicago Press, 1969.

Storer, Norman W. *The Social System of Science.* New York: Holt, Rinehart and Winston, Inc., 1966.

Survey Research Center. *The Public Impact of Science in the Mass Media.* Ann Arbor: University of Michigan Press, 1958.

Swift, Jonathan. *Gulliver's Travels*. London: Oxford University Press, 1949.

Thoenes, Piet. *The Elite in the Welfare State*. London: Faber and Faber, 1966.

Van Dyke, Vernon. *Pride and Power: The Rationale of the Space Program*. Urbana: University of Illinois Press, 1964.

Veblen, Thorstein. *The Engineers and the Price System*. New York: Viking Press, 1934.

Wallace, William K. *The Passing of Politics*. London: George Allen and Unwin, Ltd., 1924.

Ward, Lester F. *Applied Sociology*. Boston: Ginn and Company, 1906.

Wilensky, Harold L. *Organizational Intelligence: Knowledge and Policy in Government and Industry*. New York: Basic Books, 1967.

Wildavsky, Aaron. *The Politics of the Budgetary Process*. Boston: Little, Brown and Company, 1964.

ARTICLES AND PERIODICALS

Abelson, Philip H. "The President's Science Advisers," *Minerva*, III, No. 2 (Winter, 1965), 149–158.

Bock, Betty. "Relativity of Economic Evidence in Merger Cases—Emerging Decisions Force the Issue," *Michigan Law Review*, 63, No. 8 (June, 1965), 1325–1448.

Bowers, Raymond V. "The Uses of Sociology in the Military Establishment," in *The Uses of Sociology* (Edited by Paul Lazarsfeld *et al*). New York: Basic Books, 1967, 234–274.

Brodie, Bernard. "The Scientific Strategists," in *Scientists and National Policy-Making* (Edited by Robert Gilpin and Christopher Wright). New York: Columbia University Press, 1964.

Brooks, Harvey. "The Scientific Adviser," in *Scientists and National Policy-Making* (Edited by Robert Gilpin and Christopher Wright). New York: Columbia University Press, 1964.

Articles and Periodicals 309

Bryson, Lyman. "Notes on a Theory of Advice," *Political Science Quarterly*, LXVI, No. 3 (September, 1951), 321–329.

Brzezinski, Zbigniew. "Purpose and Planning in Foreign Policy," *Public Interest*, 14 (Winter, 1969), 52–73.

Davison, W. Phillips. "The Use of Sociology in Foreign Policy," in *The Uses of Sociology* (Edited by Paul Lazarsfeld et al). New York: Basic Books, 1967.

Denny, Brewster C. (ed.). "Science and Public Policy: A Symposium," *Public Administration Review*, XXVII, No. 2 (June, 1967), 95–133.

Downs, Anthony. "Some Thoughts on Giving People Economic Advice," *American Behavioral Scientist*, IX, No. 1 (September, 1965), 30–32.

DuBridge, Lee A. "Policy and the Scientists," *Foreign Affairs*, XLI, No. 3 (April, 1963), 571–588.

Dupre, J. Stefan and Gustafson, W. Eric. "Contracting for Defense: Private Firms and the Public Interest," *Political Science Quarterly*, LXXVII, No. 2 (June, 1962), 161–177.

Dupree, A. Hunter. "Central Scientific Organization in the United States Government," *Minerva*, I, No. 4 (Summer, 1963), 453–469.

Eaton, Joseph W. "Symbolic and Substantive Evaluative Research," *Administrative Science Quarterly*, VI (March, 1962), 421–442.

Eiduson, Bernice T. "Scientists as Advisors and Consultants in Washington," *Bulletin of the Atomic Scientists*, XXII, No. 8 (October, 1966), 26–31.

Fischer, Carl William. "Scientists and Statesmen: A Profile of the Organization of the President's Science Advisory Committee," in *Knowledge and Power* (Edited by Sanford A. Lakoff). New York: Free Press, 1966.

Garfinkel, Herbert. "Social Science Evidence and the School Segregation Cases," *Journal of Politics*, XXI, No. 1 (February, 1959), 37–59.

Bibliography

Glass, Bentley. "Scientists in Politics," *Bulletin of the Atomic Scientists*, XVIII, No. 5 (May, 1962), 2–7.

Goldstein, Walter. "The Science Establishment and Its Political Control," *Virginia Quarterly Review*, XLIII, No. 3 (Summer, 1967), 353–371.

Gorham, William, Drew, Elizabeth B., and Wildavsky, Aaron. "Symposium on PPBS: Its Scope and Limits," *Public Interest*, 8 (Summer, 1967), 4–48.

Grazia, Alfred de. "A Concept of Scientists and Their Organization," *American Behavioral Scientist*, VI, No. 4 (December, 1962), 30–34.

Grazia, Alfred de. "The Politics of Science and Dr. Velikovsky," *American Behavioral Scientist*, VII, No. 1 (September, 1963), 3–68.

Greenberg, Daniel S. "Mohole: The Project That Went Awry," in *Knowledge and Power* (Edited by Sanford Lakoff). New York: Free Press, 1966.

Greenberg, Daniel S. "The Myth of the Scientific Elite," *Public Interest*, I, No. 1 (Fall, 1965), 51–62.

Gustavson, W. Eric. "Science vs. Administrative Evangelism," *Public Administration Review*, XXII (Spring, 1962), 84–88.

Horowitz, Irving Louis. "Social Science Yogis and Military Commissars," *Trans-action*, V, No. 6 (May, 1968), 29–38.

Johnson, Harry G. "The Economic Approach to Social Questions," *Public Interest*, 12 (Summer, 1968), 68–79.

Jouvenal, Bertrand de. "The Political Consequences of the Rise of Science," *Bulletin of the Atomic Scientists*, XIX, No. 9 (November, 1963), 2–8.

Kash, Don E. "Is Good Science Good Politics?" *Bulletin of the Atomic Scientists*, XXI, No. 3 (March, 1965), 34–36.

Kaysen, Carl. "Model-Makers and Decision-Makers: Economists and the Policy Process," *Public Interest*, 12 (Summer, 1968), 80–95.

Lakoff, Sanford A. "The Scientific Establishment and American Pluralism," in *Knowledge and Power* (Edited by Sanford A. Lakoff). New York: Free Press, 1966.

Lakoff, Sanford A. "The Third Culture: Science in Social Thought," in *Knowledge and Power* (Edited by Sanford A. Lakoff). New York: Free Press, 1966.

Lane, Robert E. "The Decline of Politics and Ideology in a Knowledgeable Society," *American Sociological Review*, 31 (October, 1966), 649–662.

La Porte, Todd R. "Politics and Inventing the Future: Perspectives in Science and Government," *Public Administration Review*, XXVII, No. 2 (June, 1967), 95–133.

Leiserson, Avery. "Science and the Public Life," *Journal of Politics*, XXIX, No. 2 (May, 1967), 241–260.

Leiserson, Avery. "Scientists and the Policy Process," *American Political Science Review*, LIX, No. 2 (June, 1965), 408–416.

Lowi, Theodore J. "American Business, Public Policy, Case Studies, and Political Theory," *World Politics*, XVI, No. 4 (July, 1964), 677–715.

Lowi, Theodore J. "Making Democracy Safe for the World: National Politics and Foreign Policy," in *Domestic Sources of Foreign Policy* (Edited by James N. Rosenau). New York: Free Press, 1967.

MacDonald, Gordon J. F. "Science and Space Policy: How Does It Get Planned?" *Bulletin of the Atomic Scientists*, XXIII, No. 5 (May, 1967), 2–9.

Machajski, Waclaw. "On the Expropriation of the Capitalists," Selection from *The Intellectual Worker* in *The Making of Society* (Edited by V. F. Calverton). New York: Random House, 1937.

Mainzer, Lewis C. "The Scientist as Public Administrator," *Western Political Quarterly*, XVI, No. 4 (December, 1963), 814–829.

Merton, Robert K. "The Ambivalence of Scientists," in *Science and Society* (Edited by Norman Kaplan). Chicago: Rand McNally, 1965.

Merton, Robert K. "The Matthew Effect in Science," *Science,* CLIX (January 5, 1968), 56–63.

Merton, Robert K. "The Role of Applied Social Science in the Formation of Policy: A Research Memorandum," *Philosophy of Science,* XVI, No. 3 (July, 1949), 161–181.

Merton, Robert K. "Science, Technology and Society in Seventeenth Century England," *Osiris,* IV (1938), 360–362.

Mesthene, Emmanuel G. "Can Only Scientists Make Government Science Policy?" *Science,* CXLV, No. 3629 (July 17, 1964), 237–240.

Mesthene, Emmanuel G. "The Impact of Science on Public Policy," *Public Administration Review,* XXVII (June, 1967), 97–104.

Millikan, Max F. "Inquiry and Policy: The Relation of Knowledge to Action," in *The Human Meaning of the Social Sciences* (Edited by Daniel Lerner). New York: World Publishing Company, 1959, 158–182.

Moeser, John. "The Space Program and the Urban Problem: Case Studies of the Components of National Consensus," Staff Discussion Paper, Program of Policy Studies in Science and Technology, George Washington University, Spring, 1969.

Moore, Joan W. and Moore, Burton M. "The Role of the Scientific Elite in the Decision to Use the Atomic Bomb," *Social Problems,* VI, No. 1 (Summer, 1958), 78–85.

Myrdal, Gunnar. "The Relation Between Social Theory and Social Policy," *British Journal of Sociology,* 4 (1953), 210–242.

Nomad, Max. "Masters—Old and New: A Social Philosophy without Myths," in *The Making of Society* (Edited by V. F. Calverton). New York: Random House, 1937.

Olson, Manour, Jr. "Economics, Sociology and the Best of all Possible Worlds," *Public Interest,* 12 (Summer, 1968), 96–118.

Pettigrew, Thomas F. and Back, Kurt W. "Sociology in the Desegregation Process: Its Use and Disuse," in *The Uses of Sociology* (Edited by Paul Lazarsfeld *et al*). New York: Basic Books, 1967, 692–724.

Piccard, Paul J. "Scientists and Public Policy: Los Alamos, August–November, 1945," *Western Political Quarterly,* XVIII, No. 2 (June, 1965), 251–262.

Articles and Periodicals 313

Polanyi, Michael. "The Republic of Science: Its Political and Economic Theory," *Minerva,* I, No. 1 (Autumn, 1962), 54–73.

Posvar, Wesley W. "The Impact of Strategy Expertise on the National Security Policy of the United States," *Public Policy,* VIII (1964), 36–68.

Price, Derek J. de Solla. "The Science of Science," in *The Science of Science* (Edited by Goldsmith and McKay) London: Souvenir Press, 1964.

Reagan, Michael D. "The Political Structure of the Federal Reserve System," *American Political Science Review,* LV, No. 1 (March, 1961), 64–76.

Reiser, Stanley Joel. "Smoking and Health: The Congress and Causality," in *Knowledge and Power* (Edited by Sanford A. Lakoff). New York: Free Press, 1966, 293–311.

Reston, James. "The H-Bomb Decision," *New York Times,* April 8, 1954, 20.

Rossi, Peter H. "Researchers, Scholars and Policy-Makers: The Politics of Large-Scale Research," *Daedalus,* XCIII, No. 4 (Fall, 1964), 1142–1161.

Salisbury, Robert H. "The Analysis of Public Policy: A Search for Theories and Roles," in *Political Science and Public Policy* (Edited by Austin Ranney). Chicago: Markham Publishing Co., 1968.

Salisbury, Robert H. and Heinz, John P. "A Theory of Policy Analysis and Some Preliminary Applications," *American Political Science Review,* LXIV, No. 1 (June, 1970).

Sanders, Ralph. "The Autumn of Power: The Scientist in the Political Establishment," *Bulletin of the Atomic Scientists,* XXII, No. 8 (October, 1966), 22–25.

Sapolsky, Harvey M. "Science Advice for State and Local Government," *Science,* CLX (April 19, 1968), 280–284.

Sayre, Wallace S. "Scientists and American Science Policy," in *Scientists and National Policy-Making* (Edited by Robert Gilpin and Christopher Wright). New York: Columbia University Press, 1964.

Schiff, Ashley L. "Innovation and Decision Making: The Conservation of Land Resources," *Administrative Science Quarterly*, II, No. 1 (June, 1966), 1–32.

Schilling, Warner R. "The H-Bomb Decision: How to Decide Without Actually Choosing," *Political Science Quarterly*, LXXVI, No. 1 (March, 1961), 24–46.

Schilling, Warner R. "Scientists, Foreign Policy, and Politics," in *Scientists and National Policy-Making* (Edited by Robert Gilpin and Christopher Wright). New York: Columbia University Press, 1964.

Shils, Edward. "Ideology and Civility: On the Politics of the Intellectual," *Sewanee Review*, LXVI, No. 3 (July–September, 1958), 450–458.

Shils, Edward. "The Intellectuals and the Powers," *Comparative Studies in Society and History*, I, No. 1 (October, 1958), 5–22.

Shils, Edward. "Social Science and Social Policy," *Philosophy of Science*, XVI, No. 3 (July, 1949), 219–242.

Skolnikoff, Eugene B. "Birth and Death of an Idea: Research in A.I.D.," *Bulletin of the Atomic Scientists*, XXIII, No. 7 (September, 1967), 38–40.

Speyer, Edward. "The Brave New World for Scientists," *Dissent*, VIII, No. 2 (Spring, 1961), 126–136.

Stein, Herbert. "Tax Cut in Camelot," *Trans-action*, 6, No. 5 (March, 1969), 38–44.

Steinbach, H. Burr. "Scientists and Public Policy," *Bulletin of the Atomic Scientists*, XVII, No. 3 (March, 1962), 10–13.

Stockfish, J. A. and Edwards, D. J. "The Blending of Public and Private Enterprise: The SST as a Case in Point," *Public Interest*, 14 (Winter, 1969), 108–117.

Storer, Norman W. "Some Sociological Aspects of Federal Science Policy," *American Behavioral Scientist*, VI, No. 4 (December, 1962), 27–30.

Strickland, Donald A. "Scientists as Negotiators: The 1958 Geneva Conference of Experts," *Midwest Journal of Political Science*, VIII, No. 4 (November, 1964), 372–384.

Taguiri, Renato. "Value Orientations and the Relationship of Managers and Scientists," *Administrative Science Quarterly*, X, No. 1 (June, 1965), 39–51.

Tarr, David W. "Military Technology and the Policy Process," *Western Political Quarterly*, XVIII, No. 1 (March, 1965), 135–148.

Toulmin, Stephen. "The Complexity of Scientific Choice: A Stocktaking," *Minerva*, II, No. 3 (Spring, 1964), 343–359.

Toulmin, Stephen. "The Complexity of Scientific Choice, II: Culture, Overhead or Tertiary Industry," *Minerva*, IV, No. 2 (Winter, 1966), 155–169.

Uyehara, Cecil H. "Scientific Advice and the Nuclear Test Ban Treaty," in *Knowledge and Power* (Edited by Sanford A. Lakoff). New York: Free Press, 1966.

Weinberg, Alvin M. "Can Technology Replace Social Engineering?" *Bulletin of the Atomic Scientists*, XXII, No. 12 (December, 1966), 4–8.

Weinberg, Alvin M. "Criteria for Scientific Choice," in *Science and Society* (Edited by Norman Kaplan). Chicago: Rand McNally, 1965.

Wengert, Norman (ed.). "Perspectives on Government and Science," *Annals of the American Academy of Political and Social Science*, CCCXXVII (January, 1960), entire issue.

Wheeler, Harvey. "The Short and Happy Life of a Research Consultantship," *Western Political Quarterly*, 13 (September, 1960), 852–857.

Whittlesey, C. R. "Power and Influence in the Federal Reserve System," *Economica*, XXX (February, 1963), 33–44.

Wik, Reynold M. "Science and American Agriculture," in *Science and Society in the United States* (Edited by Van Tassel and Hall). Homewood, Illinois: Dorsey Press, 1966.

Wilcox, Walter W. "Social Scientists and Agricultural Policy," *Journal of Farm Economics*, XXXIV, No. 2 (May, 1952), 173–183.

Wilhelm, Sidney. "Scientific Unaccountability and Moral Accountability," in *The New Sociology* (Edited by Irving Louis Horowitz). New York: Oxford University Press, 1964.

Wilson, Francis G. "Public Opinion and the Intellectuals," *American Political Science Review*, XLVIII, No. 2 (June, 1954), 321–339.

Wohlstetter, Albert. "Scientists, Seers and Strategy," *Foreign Affairs*, XLI, No. 3 (April, 1963), 466–478.

Wohlstetter, Albert. "Strategy and the Natural Scientists," in *Scientists and National Policy-Making* (Edited by Robert Gilpin and Christopher Wright). New York: Columbia University Press, 1964.

Wood, Robert C. "Scientists and Politics: The Rise of an Apolitical Elite," in *Scientists and National Policy-Making* (Edited by Robert Gilpin and Christopher Wright). New York: Columbia University Press, 1964.

UNPUBLISHED MATERIAL

Alexander, William M. "Influences of the Atomic Scientists on Enactment of the Atomic Energy Act of 1946." Unpublished Ph.D. dissertation, University of Oregon, 1963.

Eyestone, Robert. "The Life Cycle of American Public Policies: Agriculture and Labor Policy Since 1929." Paper Presented to the 1968 Annual Meeting of the Midwest Political Science Association, at Chicago, May 2–4, 1968.

Nelson, William R. "Case Study of a Pressure Group: The Atomic Scientists." Unpublished Ph.D. dissertation, University of Colorado, 1965.

Perry, Simon Daniel. "Conflict of Expectations and 'Policy Science' Behavior." Unpublished Ph.D. dissertation, Michigan State University, 1961.

Waller, Harold Myron. "Natural Scientists and Politics: Attitudes, Orientations, and Perceptions." Unpublished Ph.D. dissertation, Georgetown University, 1968.

Index

A

ACDA (Arms Control and Disarmament Agency), 100, 110-111
Active Society, The (Etzioni), 15
Access points:
  basic, 55
  on disarmament and arms control policy, 105-106, 108
  in governmental redistributive policy arena, 90
  in regulative policy arena, 140
Acheson, Dean, 106
Ackley, Gardner, 175
Administrators:
  defined, 56
  democracy and, 16-17
  as formal position, 56
  function of, 49-50
  regulative policy arena and, 144
  scientists compared with, 27
  subservience to, 46
Advisers:
  atomic scientists as, 112
  in distributive arena, 116
  economists as, 88, 152, 159, 171-179
  as formal position, 56
  in governmental redistributive policy arena, 89
  physical scientists as, 213
  scientific strategists as, 194
AEC (Atomic Energy Commission), 103, 105-108, 208
Agency for International Development (AID), 112-114
Agricultural scientists:
  in distributive policy arena, 116-117
  foreign aid policy and, 112, 113
Agriculture, Department of (*see* Agriculture policy; Conservation policy)
Agriculture policy, 31, 73, 116, 123-126
AID (Agency for International Development), 112-114
Alvarez, Luis, 208
American Business and Public Policy (Bauer, Pool, and Dexter), 151
American Dilemma, An (Myrdal), 87
American Farm Bureau Federation, 125
American Iron and Steel Institute, 148
American Medical Association, 148, 189

320 Index

Analysis of Political Behavior, The (Laswell), 19
Antitrust policy, 32, 158-161
Antitrust suits against oil industry, 80
Applied Sociology (Ward), 17
Arena (*see* Policy arena)
Argyris, Chris, 94
Arms control (*see* Disarmament and arms control policy)
Arms Control and Disarmament Agency (ACDA), 100, 110-111
Arnold, Thurman, 159, 160
Art, Robert J., 203
Atomic bomb, 191, 200, 227*n*.
  abdicating responsibility for use of, 201-202
  initial use of, 104-105
Atomic energy, 245
  government ownership of, 77-80
  medical uses of, 269-270
  post-war domestic and international controls on, 105-106
  research funds for, 249
Atomic Energy Act (1946), 105
Atomic Energy Commission (AEC), 103, 105-108, 208
Atomic scientists (*see* Atomic bomb; Defense and deterrence policy; Disarmament and arms control policy; Hydrogen bomb; Weapons policy)
Automotive industry (*see* Transportation policy; Transportation safety policy)

B

Bacon, Francis, 20
Balance of payments policy, 149-152
Baruch Plan, 105, 106
Bauer, Raymond, 150, 151
Behavioral endogenous factors, listed, 51-55
Behavioral scientists (*see* Psychiatrists; Psychologists; Social scientists)
Behavioral technologies, as redistributive, 274-276
Bell, David, 113
Bernal, John Desmond, 16
Bethe, Hans A., 52
Bethe Panel, 108
Beyond the Ruling Class (Keller), 15
Biological warfare, secrecy surrounding, 42
Biologists, 47
  in entrepreneurial policy arena, 211
  health policy making and, 185
  pollution policy and, 145, 147
  research funds available to, 249
Birth control, 240, 272, 275
Bombs:
  atomic (*see* Atomic bomb)
  hydrogen (*see* Hydrogen bomb)
Botanists, agriculture policy and, 123-125
Boulding, Kenneth, 49, 283
Braun, Werner von, 200, 215
Brecht, Bertolt, 3, 277, 290
Brimmer, Andrew, 177
Brodie, Bernie, 193
Brookings Institution (Institute for Governmental Research), 93
Brown, Harold, 215
Brown v. *Board of Education*, 265, 266
Brownlow, Louis, 94
Brzezinski, Zbigniew, 99, 101-102
Buchanan, Scott, 284, 285
Bulletin of the Atomic Scientists, 105

Bureaucracies, secrecy and power of, 42
  (*See also* Administrators)
Burns, Arthur F., 173, 174, 176
Bush, Vannevar, 106
Business (*see* Private interests)

C

Camelot, Project, 102, 132*n*., 271
Capitalism, growth of science under, 16
Change:
  procedural and substantive, 237-238
  resistance to, 90, 92
Chapman, H. H., 157
Charles, Project, 208
Checks and balances:
  scientific truth and system of, 54
  on scientists' influence, 280-290
Chemists, 147, 185
  agriculture policy and, 123-125
Church, Frank, 110
Cities:
  obstacles to solving problems of, 84-85
  political power of, 120
  as underdeveloped region of influence, 262-265
Clark, John D., 173
Clayton Act (1914), 158
Clean Air Act (1963), 147-148
Cleaveland, Frederic, 262, 263
Coal industry (*see* Extraction policy)
Cold war:
  communal security policy arena and, 180, 181
  disarmament and arms control policy and, 103-112
  effects of, on science policy, 216
  as urgency, 43

Commission on Economy and Efficiency, 92
Commonwealth of skills, defined, 19
Communal security policy arena, 10, 35, 36, 179-209
  basic tenets of, 179-182
  defense and deterrence policy in, 32, 49, 89, 190-198
  health policy in, 32, 185-190, 249
  weapons policy in, 32, 198-209, 227-228*n*.
  weather policy in, 32, 182-185
Compensatory behavior, defined, 138
Competence, defined, 53
Competition:
  legal and economic notions of, 158-159
  workable, described, 160
Compton, Arthur, 191
Comte, Auguste, 17, 287
Conflicts, 61*n*.
  in antitrust policy, 161
  in defense and deterrence policy, 196-197
  in disarmament and arms control policy, 104-107
  in distributive policy arena, 117
  in extra-national policy arena, 96
  political, 51-53
  in pollution policy, 149, 156
  in science policy, 219-220
  in social redistributive policy arena, 81-82
  in space policy, 215
  in transportation safety policy, 143
Congress:
  atomic energy control and, 106
  as basic access point, 55
  communal security policy arena and, 188-189, 205

distributive policy arena and, 114-115, 125
entrepreneurial policy arena and, 211, 215
foreign aid policy and, 112
organization policy and, 94
regulative policy arena and, 138, 139, 141, 142, 144, 145, 147, 148, 155
role of, 7
self-regulative policy arena and, 78, 80
as underdeveloped region for influence, 259-262
Consciousness, assessment of technology and need for, 273-274
Consensus politics, 84, 148
Conservation policy, 32, 153-158
Contraception, 240, 272, 275
Control:
of atomic energy, 105-106
research and political, 49
Cortner, Richard C., 266
Council of Economic Advisers, 88, 152, 171, 173-178
Council on Environmental Quality, 145
Council of International Affairs, 102
Council of Social Advisers, 88
Courts:
as basic access point, 55
as underdeveloped region of influence, 265-268
Crisis, defined, 39
Criticism dampened by secrecy, 42

D

Daddario, Emilio Q., 261, 272
Darwin, Charles, 87
Decentralization, need for scientists and, 39

Decision making:
indecision and, 43-45
science and, 144
(*See also* Science)
Decision systems:
fragmented, 10, 115, 238
integrated, 10
influence and, 238-240
regulative activities as, 139
(*See also* Policy making)
Defense, Department of (*see* Defense and deterrence policy; Foreign political policy; Weapons policy)
Defense and deterrence policy, 32, 49, 89, 190-198
(*See also* Disarmament and arms control policy; Weapons policy)
Demand patterns:
effects of, 11
exogenous factors representing, 41-46
fragmented, 10
in distributive policy arena, 115
of minorities, 81
in regulative policy arena, 139
integrated, 10
increased influence through, 238-239
political, 67
Democracy:
danger to, 16-17
growth of science in, 278-280
Democratic values, 17, 245
Depletion allowance, oil, 80
Deterrence:
development of, 106-107
mutual, 107-112
(*See also* Defense and deterrence policy)
Developing countries, science in, 291*n*.

Developing regions of influence, 268-269
Development (*see* Research and development)
Dewey, John, 95
Dexter, Lewis A., 150, 151
Diplomacy, as redistributive, 95
Disarmament and arms control policy, 31, 73, 95, 99, 103-112
(*See also* Defense and deterrence policy; Weapons policy)
Discovery, as political act, 29
Distributive policy arena:
agriculture policy in, 123-126
characteristics of, 114-115
defined, 34, 36
factors leading to, 10
transportation policy in, 31-32, 73, 117-123
Douglas, Paul, 152
Downs, Antony, 62n., 92
Dror, Yehezkel, 238, 288, 289
Dubos, Rene, 99
Dubridge, Lee, 192, 208
Dulles, John Foster, 104, 107, 192

E

East River, Project, 208
Eaton, Joseph W., 279
Eberstadt Report (1948), 93
Ecologists, pollution policy and, 145, 149
Economic factors (*see* Private interests; Socio-economic factors; Vested interests)
Economic indicators, 88
Economic management policy arena, 35, 36, 169-179, 239
basic tenets of, 169-172
fiscal and monetary policy in, 32, 172-179

Economic power affected by pollution control policy, 146
Economic Report, 88
Economists:
as advisers, 88, 152, 159, 171-179
agricultural policy and, 123-126
antitrust policy and, 158
in defense and deterrence policy making, 195
in economic management policy arena, 171-178
in extra-national policy arena, 97
foreign aid policy and, 112-114
functions of, 62n.
governmental redistribution policy arena and, 89, 90
in regulative policy arena, 140, 146, 147, 151, 152
as scientific strategists, 49
in self-regulative policy arena, 77, 78, 80
social policy and, 83
transportation policy and, 117, 119-122
in weapons policy making, 202
Economy, as factor in organization policy, 92-94
Edelman, Murray, 77
Education, technological solutions for, 253n.
Educational policy, as distributive, 115
Edwards, D. J., 122
Efficiency, as factor in organization policy, 92, 94
Einstein, Albert, 105, 200, 201
Eisenhower, Dwight D., 13, 107, 108, 151-152, 173, 174, 178, 214
Elite in the Welfare State, The (Thoenes), 16
Elites:
illegal, 16-17

interdependence of, 283
scientific, 284, 293*n*.
  notion attacked, 46
strategic, 15
of totalitarian systems, 277-278
Endogenous factors, 46-55, 58
Engineers:
  in distributive policy arena, 117
  in extra-national policy arena, 97
  extraction and resource, 78
  influence of, 47
  pollution policy and, 147
  systems, 49, 50, 264
  transportation policy and, 118-119
  transportation safety policy and, 142
Engineers and the Price System, The (Veblen), 18
Engler, Robert, 79
Enthoven, Alain, 195
Entrepreneurial policy arena, 10, 35, 36, 209-220
  basic tenets of, 209-212
  distributive policy compared with, 115
  science policy in, 32-33, 216-220, 229*n*.-231*n*.
  space policy in, 32, 84, 212-216, 249
Environmental policy (*see* Agriculture policy; Conservation policy; Extraction policy; Health policy; Pollution policy; Transportation policy)
Estate, scientists as, 18
Etzioni, Amitai, 15, 288
Executive Branch:
  as basic access point, 55
  indecision in, 43-45
  influence and policy arenas in, 63-70
  (*See also* Influence; Policy arena)
  policy making activities in (1945-1968), 6-11, 25, 59*n*.-60*n*.
  (*See also* Policy; Policy making)
Exogenous factors, 41-46, 58
Expertise, non-scientific, defined, 51
Extra-national policy arena, 10, 35, 36, 95-114
  disarmament and arms control policy in, 31, 73, 95, 99, 103-112
  foreign aid policy in, 31, 73, 95, 112-114
  foreign political policy in, 31, 73, 98-103, 131*n*.
  redistributive nature of, 95-96
Extraction policy, 31, 73, 78-81
Eyestone, Robert, 10, 34, 125

F

Factors:
  economic and socio-economic, 16, 81-89, 243-245
  (*See also* Private interests)
  endogenous, 46-55, 58
  exogenous, 41-46, 58
  participational, 55-58
  political: of defense and deterrence policy, 191-193, 197
    in disarmament and arms control policy, 109-112
    in distributive policy arena, 119, 120, 124, 126
    exogenous factors as, 41-46
    in fiscal and monetary policy, 173, 174
    in foreign political policy, 102
    in health policy, 186, 187

importance of, 243-245
of regulative policy arena, 138, 147, 155, 157-159
of science policy, 217-219
in scientists' relation to policy making, 9-10
in self-regulative policy arena, 7, 75
of social redistributive policy arena, 85-88
of space policy, 214, 215
of weather policy making, 184
producing self-regulation, 10
Fallout, 107-112
Federal Council on Science and Technology, 217
Federation of Atomic Scientists, 105
Ferkiss, Victor, 287
Fermi, Enrico, 191, 208
Feuer, Lewis, 202
Fields, defined, 47
Fieser, Louis F., 201
Fire and Water: Scientific Heresy in the Forest Service (Schiff), 156, 157
Fiscal and monetary policy, 32, 172-179
Fixes, technological, 84, 241
Flash, Edward, Jr., 176
Flow of information, secrecy and, 42
Fluoridation, 270
Foreign aid policy, 31, 73, 95, 112-114
Foreign political policy, 31, 73, 98-103, 131n.
Formal positions:
access points and, 55-56
degree of control on, 62n.
held by economists, 172
in space policy making, 213
in transportation safety policy, 143

Foster, John S., 109
Fragmented decision systems, 10, 115, 238
Fragmented demands, 10
in distributive policy arena, 115
of minorities, 81
regulation and, 139, 239
Franck, James, 104
Franck Committee, 53, 104, 191
Franklin, Benjamin, 183
Fromm, Erich, 194
Full Employment Act (1946), 89, 171, 172
Full Opportunity and Social Accounting Act (proposed 1967), 89
Functional stages of participation, 56-57, 62n.

G

Galbraith, John Kenneth, 170, 178
Galileo, 290
Galileo (Brecht), 3, 279
Game theory, 49, 195, 196
Garrison (warfare) state, 98, 110, 190
Genetic code, altering, 273
Geneticists, 265, 267
Geneva Conference of Experts (1958), 108, 109
Ghetto disturbances:
as urban problems, 84
as urgency, 43
Gilpin, Robert, 52
Government (*see* Congress; Executive Branch)
Governmental redistributive policy arena, 10, 35, 36, 73, 89-95
characterized, 90
issues involved, 89
organization policy in, 31, 73, 91-95, 237-238
Greenberg, Daniel S., 46, 219

Greenfield, Meg, 215
Gross, Bertram, 89
Gulick, Luther, 94
Gulliver's Travels (Swift), 18

H

Hamilton, Fowler, 113
Harriman, Averell, 282
Harrington, Michael, 83, 87
Harris, Fred, 260
Harrisson, Wilfrid, 33
Hauge, Gabriel, 151
Health, Education and Welfare, Department of (*see* Education policy; Health policy; Social policy; Transportation policy)
Health policy, 32, 185-190, 249
Heinz, John, 10, 115
Heller, Walter, 174-176
High-influence policy process, 8, 12, 33, 167-231
　characterized, 167-169
　in communal security arena, 10, 35, 36, 179-209
　　basic tenets of, 179-182
　　defense and deterrence policy in, 32, 49, 89, 190-198
　　health policy in, 32, 185-190, 249
　　weapons policy in, 32, 198-209, 227-228n.
　　weather policy in, 32, 182-185
　in economic management policy arena, 169-179
　　basic tenets of, 169-172
　　fiscal and monetary policy, 172-179
　in entrepreneurial policy arena, 10, 35, 36, 209-220
　　basic tenets of, 209-212
　　distributive policy compared with, 115
　　science policy in, 32-33, 216-220, 229n.-231n.
　　space policy in, 32, 84, 212-216, 249
　　table of, 36
Hindsight, Project, 204, 227n.
Hines, Howard M., 86
Hitch, Charles J., 193, 195
Hoover Commission on the Organization of the Executive Branch of the Government, 90-93
Housing policy, as distributive, 115
Humphrey, George, 174
Huntington, Samuel, 89, 197-198
Hydrogen bomb, 192, 200
　development of, 106-107, 208

I

Incrementalism:
　in social redistributive policy arena, 82
　in weapons policy process, 205-206
Indecision among leadership, defined, 43-44
Industry, influence of, 7
　(*See also* Private interests)
Influence:
　checks and balance on, 280-290
　defined, 28-30
　developing regions of, 268-269
　factors shaping, 40-58
　　endogenous factors, 46-55, 58
　　exogenous factors, 41-47, 58
　　participational factors, 55-58
　　political factors, 41-46, 243-245

socio-economic factors, 81-89, 243-245
implications of, 246-251
levels of, defined, 40
 (*See also* High-influence policy process; Low-influence policy process; Moderate-influence policy process)
policy arenas and, 63-70
 (*See also* specific policy arenas)
policy-making conditions affecting, 234-242
underdeveloped regions of, 258-268
 Congress, 259-262
 courts and legal system, 265-268
 states and cities, 262-265
Inside Bureaucracy (Downs), 92
Institute for Governmental Research (Brookings Institution), 93
Integrated decision systems, 10
 influence and, 238-240
 regulative activities as, 139
Integrated demand patterns, 10
 increased influence through, 238-239
Intellectuals:
 role of, 17
 scientists compared with, 26-27
Intelligence and Research and Policy Planning Council, Bureau of, 99
Interest groups (*see* Private interests; Vested interests)
Interim Committee, 104, 191
Interior, Department of (*see* Conservation policy; Extraction policy)
International relations modified by atomic bomb, 104
International Scientific and Technological Affairs, Office of, 99
Interstate Oil Compact Commission, 76, 78
Intrusion of privacy, technology and, 270, 271
Intuition in foreign political policy, 101
Iowa Oleo case, 126, 135*n*.
Issue area, 10
Issues created by technology, 269-272

J

Jarolem, Stanley, 203
Jefferson, Thomas, 183
Johnson, Lyndon B., 175, 177
Journalists, scientists compared with, 27
Justice, Department of (*see* Antitrust policy; Courts; Extraction policy; Legal system)

K

Kahn, Daniel, 203
Kahn, Herman, 193, 196
Kaplan, Abraham, 28
Keller, Suzanne, 15
Kennedy, John F., 110, 150, 151, 174-176, 191, 194, 198, 214, 215, 282
Kennedy Trade Expansion Act, 150-151
Keynes, John Maynard, 87, 174
Keyserling, Leon, 173, 176
Killian, James, 107
Kistiakowsky, George, 107, 206
Knowledge:
 policy-issue, 289
 power and, 221*n*.
 unit of, defined, 15-16
 (*See also* Science)

L

Labor policy, 115
Labor problems, technology and, 270
Lapp, Ralph, 45, 194, 204-205
Lasswell, Harold, 17, 19, 28, 56
Latter, Albert 108
Law and order, as response to social problems, 88
Lawrence, Ernest D., 106, 191, 208
Lawyers:
  antitrust policy and, 158, 159
  scientists compared with, 26-27
Leaders:
  political: analyzing personalities of, 17
    communal security policy arena and, 199, 201, 206-207
    comprehension of science by, 50-51
    extra-national policy arena and, 97
    government redistributive policy arena and, 90
    indecision among, 43-44
    orientation of, 45-46, 50-51
    regulative policy arena and, 141-143, 145, 155-156
    scientists compared with, 26-27
    scientists as useful to, 18
    self-regulative policy arena and, 74-75
    in totalitarian systems, 277-278
    (See also Executive Branch; Policy making; Political leaders)
  scientists as (see Scientists)
Legal system, as underdeveloped region of influence, 265-268
Legislative Evaluation for the Congress, Office of, 89
Lewin, Kurt, 95
Life styles:
  as constraints on health policy, 186-187
  mental health and, 223n.
Lincoln, Project, 208
Literature on scientists' relationship to political power, table, 14
Low-influence policy process, 8, 35
  in distributive policy arena: agriculture policy in, 123-126
    characteristics of, 114-115
    defined, 34, 36
    factors leading to, 10
    transportation policy in, 31-32, 73, 117-123
  in extra-national policy arena, 10, 35, 36, 95-114
    disarmament and arms control policy in, 31, 73, 95, 99, 103-112
    foreign aid policy in, 31, 73, 95, 112-114
    foreign policy arena in, 31, 73, 98-103, 131n.
  in governmental redistributive policy arena, 10, 35, 36, 73, 89-95
    characterized, 90
    issues involved, 89
    organization policy in, 31, 73, 91-95, 237-238
  in self-regulative policy arena, 18, 35, 36, 73, 75-81
    defined, 34, 36
    distributive policy arena compared with, 115
    extraction policy in, 31, 73, 78-81
    issues involved in, 75

private interests and, 76-78
social redistributive policy arena compared with, 83-84
in social redistributive policy arena, 10, 35, 36, 73, 81-89
nature of conflicts in, 81-82
social policy in, 31, 73, 85-89
urban problems and, 84-85
table of, 36
Lowi, Theodore, 9, 10, 34, 35, 37, 38, 81, 125
Lysenko, Trofim D., 278

M

McConnell, Grant, 154
McMahon Act, 105
McNamara, Robert S., 194, 195, 198, 202, 203, 205, 215
Magnuson, Warren, 86
Manhattan Project, 53
Manufacturing Chemists Association, 148
Marx, Karl, 16, 87
Mass media (see Media)
Mass transit system, obstacles to development of, 118-121
Massive retaliation doctrine, 104, 107
Massocracy, defined, 277
Mathematicians, as scientific strategists, 49
May-Johnson Bill, 105, 106
Mayo, Elton, 95
Media, use of, 62n.
  atomic energy control and, 105-106
  for communal security policies, 182
  for conservation policy, 156
  disarmament and arms control policy and, 97, 109-111
  fiscal and monetary policy making and, 176-177
  for pollution policy, 145
  for regulative policy arena, 140
Medical profession, 241
  atomic energy used by, 269-270
  health policy and, 185, 189
Mental health, life styles and, 223n.
Merriam, Charles, 94
Merton, Robert, 57
Metapolicymaking, defined, 238
Meteorologists, 183, 185
Military-industrial-political-scientific complex, 98
  coinciding objectives of, 44-45
Military orientation, secrecy and, 45
Military policy (see Defense and deterrence policy; Weapons policy)
Millikan, Max F., 87
Minerals industry (see Extraction policy)
Moderate-influence policy process, 8, 11, 35
  table of, 36
  (See also Regulative policy arena)
Moderately high level of influence (see High-influence policy process)
Moderately low level of influence (see Low-influence policy process)
Moeser, John, 84
Mohole, Project, 229n., 260
Mondale, Walter, 88
Monopoly, legal and economic notions of, 158-159
Moscow Treaty (1963), 108, 109
Moynihan Report, 87

Myrdal, Gunnar, 87

N

Nader, Ralph, 142-144, 188
"Nader's Raiders," 142
NASA (National Aeronautics and Space Administration), 212-215
National Academy of Engineering, 149
National Academy of Sciences, 149, 217
National Aeronautics and Space Administration (NASA), 212-215
National Bureau of Economic Research, 93
National Coal Association, 148
National Environmental Policy Act (1970), 145
National Institutes of Health (NIH), 188
National Records Management Council, 93
National Science Foundation, 86, 217
National security:
 atomic energy and, 80
 requirements of, 41
Needs fulfiilled through science, 257
Nelson, Gaylord, 144
New Atlantis, The (Bacon), 20
New Economies, 82
New Industrial State, The (Galbraith), 173, 178
New Republic (magazine), 203
NIH (National Institutes of Health), 188
Nixon, Richard M., 189, 202
Non-scientific expertise, defined, 51
Non-zero sum situations, 68
 communal security policy arena as, 180
 (*See also* Communal security policy arena)
 distributive policy arena as, 114
 (*See also* Distributive policy arena)
 influence increased in, 234-236
 (*See also* Entrepreneurial policy arena)
Nourse, Edwin, 173

O

Office of Science and Technology, (OST), 55, 103, 217
Oil and Gas, Office of, 80
Oil industry (*see* Extraction policy)
Okun, Arthur, 175
Oppenheimer, J. Robert, 106, 107, 191, 192, 208
Organization policy, 31, 73, 91-95, 237-238
OST (Office of Science and Technology), 55, 103, 217
Other America, The (Harrington), 83, 87

P

Parsons, Talcott, 15
Participation:
 functional stages of, 56-57n., 62n.
 patterns of, 67
Participational factors, 55-58
Passing of Politics, The (Wallace), 16
Pauling, Linus, 52, 109, 194
PCAM (President's Committee on Administrative Management), 93-94
Peck, Merton J., 205
Personnel policy, as distributive, 115

Physical scientists:
  in defense and deterrence policy making, 191-192
  in entrepreneurial policy arena, 211-214
  image of, 245
  pollution policy and, 145, 146
  science policy and, 219
  technologies developed by, 240-242
Physical technologies, as redistributive, 274
Physicists, 47
  foreign political policy and, 100
Pinchot, Gifford, 155
Planning, hypothetical nature of, 49
Planning-Programming-Budgeting System (PPBS), 93, 195, 225n., 247
Plato, 13
PMA (Production and Marketing Association), 124
Policy:
  categorized, 11
  defined, 30
  types of: policy arenas, policy process and, table, 9
    studied, 31-33
  (See also specific type of policy)
Policy arena:
  defined, 9-10
  degree of development of, 43
  types of, 34-39
  (See also specific type of policy arena)
Policy makers, administrators compared with, 56
  (See also Administrators; Executive Branch; Leaders)
Policy making:
  executive, 6-11, 25, 59n.-60n.
  (See also Administrators; Executive Branch; Leaders)
  influence affected by conditions of, 234-242
  (See also Influence)
  outcomes of, 12-13
  process of: defined, 33-34
    policy types, policy arena and, table, 9
    (See also Policy; Policy arena)
    underdevelopment of, 43
    (See also High-influence policy process; Low-influence policy process; Moderate-influence policy process)
  scientific component of, defined, 48
  scientists' relationship to, 4-6, 11-12
    concepts on, 15-20
    literature on, 13-14
    political factors and, 9-10
    variables in, 6-8
    (See also Science; Scientists)
Political conflicts, example of, 51-53
Political demands, effects of, 67
Political factors:
  of defense and deterrence policy, 191-193, 197
  in disarmament and arms control policy, 109-112
  in distributive policy arena, 119, 120, 124, 126
  exogenus factors as, 41-46
  in fiscal and monetary policy, 173, 174
  in foreign political policy, 102
  in health policy, 186, 187
  importance of, 243-245
  of regulative policy arena, 138, 147, 150-151, 155, 157-159
  of science policy, 217-219

332 Index

in scientists' relation to policy making, 9-10
in self-regulative policy arena, 7, 75
of social redistributive policy arena, 85-88
of space policy, 214, 215
of weather policy making, 184
Political leaders:
analyzing personalities of, 17
communal security policy arena and, 199, 201, 206-207
comprehension of science by, 50-51
extra-national policy arena and, 97
government redistributive policy arena and, 90
indecision among, 43-44
orientation of, 45-46, 50-51
regulative policy arena and, 141-143, 145, 155-156
scientists compared with, 26-27
scientists as useful to, 18
self-regulative policy arena and, 74-75
in totalitarian systems, 277-278
(*See also* Congress; Executive Branch; Policy making)
Politics:
consensus, 84, 148
science and, 227-280
social sciences and, 85-88
(*See also* Checks and balance; Political factors; Political leaders)
Politics of Oil, The (Engler), 79
Politics, Pressure and the Tariff (Schattschneider), 150
Pollack, Herman, 93
Pollution, technology creating, 269-270, 272-274
Pollution policy, 83, 115, 145-149, 164n.

Pool, Ithiel de Sola, 150, 151
Population policy:
as redistributive, 81
in self-regulative policy arena, 77
Positions (*see* Formal positions)
Power:
change as threat to, 90, 92, 225n.
desire for, 46, 53-54, 293n.
economic, pollution policy and, 146
(*See also* Private interests)
knowledge and, 221n.
urban political, 120
(*See also* Politics)
PPBS (Planning-Programming-Budgeting System), 93, 195, 225n., 247
Prerequisites to influence, 68
Presidency, as basic access point, 55
(*See also* Executive Branch)
President's Committee on Administration Management (PCAM), 93-94
President's Science Advisor, 55, 217
President's Science Advisory Committee (PSAC), 55, 102-103, 107, 108, 147, 205, 217
Price, Don K., 18, 47, 49, 206
Private enterprise (*see* Private interests)
Private interests:
as beneficiaries of distributive policy arena, 114-115
(*See also* Distributive policy arena)
congeniality to, defined, 44-45
in defense and deterrence policy making, 194-195
in entrepreneurial policy arena, 209-210, 212

in foreign aid policy, 113-114
in governmental redistributive policy arena and, 91
in health policy, 187-188
in regulative policy arena, 138-139, 141
  antitrust policy, 158-161
  automobile industry, 141, 142, 144, 145
  conservative policy, 153-156
  pollution policy, 145-149
  trade policy, 150-151
in self-regulative policy arena, 76-78
in weapons policy, 200, 204
(*See also* Vested interests)
Procedural change, 237-238
Production and Marketing Association (PMA), 124
PSAC (President's Science Advisory Committee), 55, 102-103, 107, 108, 147, 205, 217
Psychiatrists, 17, 265, 267
Psychologists, 49
Public Interest, The (Stockfish and Edwards), 122
Public opinion, molding, 62*n*.
(*See also* Media)

Q

Quarles, Francis, 108

R

Rabi, I. I., 107, 208
Rabinowitch, Eugene, 104
Railroads (*see* Transportation safety policy)
RAND Corporation, 108, 111, 196, 264
Rationality, 294*n*.
  technical, defined, 287

R&D (*see* Research and development)
Reagan, Michael D., 248, 249
Redistributive policy arena, defined, 34, 36
(*See also* Governmental redistributive policy arena; Social redistributive policy arena)
Redistributive technologies, 274-276
Regulation:
  defined, 137-138
  as disguised distribution, 116
  fragmented demand pattern and, 139, 239
  political nature of, 144
Regulative policy arena, 10, 137-165, 236
  antitrust policy and, 32, 158-161
  characterized, 138-141
  conservation policy in, 32, 153-158
  defined, 34, 36
  pollution policy in, 32, 115, 145-149, 164*n*.
  trade policy, 115-116, 149-152
  transportation safety policy in, 31, 121, 141-145
Regulatory agencies in transportation policy, 119-123
Reorganization, nature of, 90
Research and development (R&D):
  for defense and deterrence policy, 195
  freedom of, 250-251
  funds for, 61*n*., 244, 245
    agricultural, 116, 124
    annual budget, 47
    atomic energy, 249
    Congress and, 260
    for disarmament and arms control, 99, 111
    effects of support, 67

on foreign aid problems, 113
in foreign political policy, 99, 102
importance of, 35-36, 248
from states and cities, 262
oil industry, 79
political control and, 49
on transportation safety, 142, 143
Research organizations, as basic access point, 55
Researchers, as formal positions, 56
Responsibility of scientists, 20
Reuss, Henry, 261
Ribicoff, Abraham, 89, 144
Ricardo, David, 87
Robinson, James A., 30
Rockefeller, Nelson, 282
Rockefeller Panel, 93
Roosevelt, Franklin D., 93, 105, 125, 200, 201
Roosevelt, Theodore, 155
Rosenau, James, 10

S

Saint-Simon, Comte de (Claude Henri de Rouvroy), 18
Salisbury, Robert, 10, 34, 77, 115
SALT (Strategic Arms Limitation Talks), 109
Sapolsky, Harvey M., 264
Sayre, Wallace, 126
Schattschneider, Elmer, 150
Schelling, Thomas, 111, 193, 196
Scherer, Frederic M., 205
Schiff, Ashley, 156, 279
Schilling, Warner, 46
Science:
 defined, 5-6, 8
 difficulty in comprehension of, 50-51, 260
 function of, in policy making, 5-6
 hypothetical character of, 47
 technology and, 27-28
  (*See also* Technology)
  (*See also* Expertise; Knowledge; Scientists; Skill)
Science and State Government (Cleaveland), 262
Science policy, 32-33, 216-220, 229*n*.-231*n*.
Science-related activities in foreign political policy, 100, 131*n*.
Science, Scientists and Politics (Buchanan), 284
Scientific elites, 284, 293*n*.
 notion of, attacked, 46
Scientific Estate, The (Price), 18, 54
Scientific field, as substantive endogenous factor, 47
Scientific strategists:
 defense and deterrence policy making and, 191-197
 emerging breed of, 49, 192
 in symbolic activities, 240
 systems used by, 196
Scientists:
 autonomous, 18-19
 defined, 26-27
 dilemma facing, 3-5
 relationship of, with Congress, 259-262
  (*See also* Congress)
 as skill group, 19-20
 usefulness of, to society, 15-18
  (*See also* Society)
  (*See also* Elites; Influence; Policy; Policy arena; Policy making; *and* specific types of scientists)
SCS (Soil Conservation Service), 124
Secrecy:
 hostility toward, 62*n*.
 in Pentagon, 45

shortcomings of, 41-42
of weapons policy, 199, 207
Security:
   national: atomic energy and, 80
      requirements of, 41
   weapons policy and need for, 207-209
Security state, 170, 279-280
Self-consciousness of society, social scientists and, 245
Self-regulation:
   factors producing, 10
   in science policy, 218
Self-regulative policy arena, 18, 35, 36, 73, 75-81
   defined, 34, 36
   distributive policy arena compared with, 115
   extraction policy in, 31, 73, 78-81
   issues involved in, 75
   private interests and, 76-78
   social redistributive policy arena compared with, 83-84
Selznick, Philip, 94
Sex, "changing" individual's, 271-273
Sherman Antitrust Act (1890), 158
Simon, Herbert, 95
Simulation, 195, 196
Situation-defining role in regulative policy arena, 140-141
Skill group, scientists as, 19-20
Skolnikoff, Eugene, 100, 101
Smith, Adam, 195, 266
Smith, Cyril, 208
Smith, James Ward, 281
Snow, C. P., 50, 53
Social classes in social redistributive policy arena, 81-83
Social Darwinism, 17
Social Function of Science, The (Bernal), 16
Social indicators, 88, 89
Social Insecurity (Steiner), 88
Social institution, science as, 277
Social policy, 31, 73, 85-89
Social redistributive policy arena, 10, 35, 36, 73, 81-89
   nature of conflicts in, 81-82
   social policy in, 31, 73, 85-89
   urban problems and, 84-85
Social Report, proposals for, 88
Social scientists:
   agricultural policy and, 123
   dilemma of, 245-246
   in entrepreneurial policy arena, 211
   in extra-national policy arena, 97, 100-102
   governmental redistributive policy arena and, 89-95
   image of, 245
   involved in Congressional reform, 259
   pollution policy and, 146
   redistributive policy arena and, 82-89
   solutions proposed by, 241, 242
   in symbolic activities, 240
   view held on, 248-249
   (*See also* Economists)
Socialism, science and, 16
Society:
   effects of technology on, 269-272
      need to anticipate problems, 272-274
   scientists and: checks and balance system and, 280-290
      interdependence, 257-258
      political systems and, 276-279, 291n.-292n.
      scientists as functionally useful to, 15-18
      society's commitment to, 243-245

Socio-economic factors, 81-89, 243-245
Sociologists (see Social scientists)
Soil Conservation Service (SCS), 124
Soviet Union (USSR), 101
 defense and deterrence policy making and, 192
 disarmament and arms control policy and, 106-109
 effect of, on space policy, 212-215
 as totalitarian state, 278
Space policy, 32, 84, 212-216, 249
Specialization, degree of, defined, 47
Spencer, Herbert, 17, 87, 266
SST (see Supersonic transport jet)
State, Department of (see Disarmament and arms control policy; Foreign aid policy; Foreign political policy)
States, as underdeveloped regions of influence, 262-265
Statisticians in self-regulative policy arena, 18
Stein, Herbert, 174, 175
Steiner, Gilbert, 88
Stimson, Henry, 53
Stockfish, J. A., 122
Strategic Arms Limitation Talks (SALT), 109
Strategic conflicts, examples of, 52
Strategic elites, defined, 15
Strategists (see Scientific strategists)
Strategy of Conflict (Schelling), 196
Strauss, Lewis L., 108
Substantive change, 237-238
Substantive endogenous factors, listed and defined, 47-51, 58
Supersonic transport jet (SST), 115, 120
 case study on, 121-123
Support:
 in communal security policy arena, 182, 184, 186, 199, 239
 for conservation policy, 155
 degree of, defined, 42
 effects of, 67
 for entrepreneurial policy arena, 210, 211, 214, 216
 in fiscal and monetary policy making, 179, 239
 for foreign aid policy, 113
 in governmental redistributive policy arena, 91
 in regulative policy arena, 140, 146-147, 149, 150, 152
 for social redistributive policy arena, 84
Swift, Jonathan, 18, 19
Symbolic Use of Politics, The (Edelman), 77
Système de Politique Positive (Comte), 287
Systems analysts, 47
 in defense and deterrence policy making, 193-194
 government redistributive policy arena and, 89, 90
 transportation policy and, 117-120
 weapons policy making and, 202
Systems engineers, 49, 50, 264
Szilard, Leo, 104, 201

T

Taft, William Howard, 92

Index

Tariff policy, as distributive, 115-116
Tarr, David, 206
Team research, 250
Technical rationality, 294n.
 defined, 287
Technological fixes, attraction of, 84, 241
Technological Man: The Myth and Reality (Ferkiss), 287
Technologies, behavioral, 274-276
Technology:
 to combat pollution, 146
 to detect underground testing, 108-109
 issues created by, 269-272
 need to anticipate, 272-274
 new automotive, 120
 physical, 274
 prestige rescued through, 213
 for regulative purposes, 140
 science and, 27-28
 (*See also* Science)
Teller, Edward, 52, 53, 106-109, 200, 208
Test bans, 107-112
TFX airplane, 202-203
Thermonuclear War, On (Kahn), 196
Thoenes, Piet, 16, 46
Tizard, Sir Henry, 200
Tobacco Institute, 188
Totalitarian systems, 277-278
Toward A Social Report (H.E.W.), 88
Trade Expansion Act (1963), 150-151
Trade policy, 115-116, 149-152
Transportation, Department of (*see* Transportation policy)
Transportation policy, 31-32, 73, 117-123
Transportation safety policy, 31, 121, 141-145
Truman, Harry S, 173, 208

U

Underdeveloped regions of influence, 258-268
 Congress as, 259-262
 courts and legal system as, 265-268
 states and cities as, 262-265
Underdevelopment of policy-making process, defined, 43
Urban problems, 84-85, 264
Urban renewal policy, nature of, 83, 115
Urgency:
 in communal security arena, 182
 defined, 42-43
 in disarmament and arms control policy, 111
 in extra-national arena, 98
 in self-regulative policy arena, 78
 in social redistributive policy arena, 84
USSR (*see* Soviet Union)

V

Valuational conflicts, examples of, 52
Values:
 in communal security policy arena, 179-180, 183, 185-186, 199
 democratic, 17, 245
 in entrepreneurial policy arena, 209, 212, 217
 in extra-national policy arena, 95, 98
 in governmental redistributive policy arena, 89
 in regulative policy arena, 150, 153, 158
 in social redistributive policy arena, 81

Veblen, Thorstein, 17, 18
Verne, Jules, 213
Vested interests:
  in communal security policy arena, 181, 186, 194, 204, 239
  defined, 54-55
  in distributive policy arena, 117
  in entrepreneurial policy arena, 216, 242
  in extra-national policy arena, 98
  in governmental redistributive policy arena, 91
  in transportation policy, 118
  (*See also* Private interests)
Visibility:
  extent of, defined, 41-42
  in self-regulative policy arena, 75
  (*See also* Secrecy)
Vista, Project, 208
Von Braun, Werner, 200, 215
Von Neumann, John, 200

W

Wallace, William K., 16
Ward, Lester, 17-18
Warfare, biological, secrecy surrounding, 42
Warfare (garrison) state, 98, 110, 190
Waste problem, 270
Water Pollution Control Act (1948), 148
Water resources policy, 115
Weapons culture, 45, 110, 204-205
Weapons policy, 32, 198-209, 227*n*.-228*n*.
  (*See also* Defense and deterrence policy; Disarmament and arms control policy)
Weather modification programs, 184-185
Weather policy, 32, 182-185
Webb, James, 215
Weber, Max, 42
Weiner, Norbert, 4
Weisner, Jerome, 107, 110, 215
Welfare policy, 81, 83, 88
Welfare state, 16-17
Wigner, Eugene, 52, 201
Willoughby, Frank, 93
Wohlstetter, Albert, 196
Wood, Robert C., 6, 19-20
Workable competition, defined, 160

Z

Zacharias, Jerrold R., 107
Zero-sum situations, 68
  extra-national policy arena as, 95
  (*See also* Extra-national policy arena)
  governmental redistributive policy arena as, 90
  (*See also* Governmental redistributive policy arena)
  influence decreased in, 234-236
  regulative policy arena as, 139
  (*See also* Regulative policy arena)
Zoologists, 123-125
Zuckerman, Sir Solly, 218